Among the Marvelous Things

The Media of Social Communications and the Next Generation of Pastoral Ministers

Edited by Dr. Sebastian Mahfood, OP

En Route Books and Media, LLC

St. Louis, MO

⚓ENROUTE
Make the time

En Route Books and Media, LLC
5705 Rhodes Avenue
St. Louis, MO 63109

Cover credit: TJ Burdick

Library of Congress Control Number:
2020930061

ISBN-13: 978-1-950108-84-8 and
978-1-950108-97-8

DEDICATION

This book is dedicated to Saint Francis de Sales, Patron Saint of Journalists, upon whose assistance we call in this new decade, the third of the third millennium, in our establishment of communications plans in all programs of pastoral formation that seek to prepare future ministers for service to the Church and to the world.

ACKNOWLEDGMENTS

The general editor would like to thank Msgr. Peter Vaccari, Rector of Saint Joseph's Seminary in the Archdiocese of New York from 2012 till 2019 and wish him many continued blessings and growth in his ministry as he departs from Yonkers to lead the Catholic Near East Welfare Association.

It was through Msgr. Vaccari's initiative in the fall of 2019 to train the seminarians and lay ministers in the theology of social communications that this book came into being.

CONTENTS

About this Book i

Preface by Sister Marysia Weber, RSM, DO, MA. iii
Foreword by Dr. Timothy Lock ix

Introduction by Dr. Sebastian Mahfood, OP 1

Part I: Selections from St. Joseph's Seminary – Dunwoodie

    Presentation 1 by Joan Brisson.. 11
    Presentation 2 by Tony George....................................... 23
    Presentation 3 by Steven McClernon 35
    Presentation 4 by Jinwoo Nam....................................... 45
    Presentation 5 by Tobechukwu Offiah 59
    Presentation 6 by Anh Ngoc-Quoc Vu............................ 71
    Presentation 7 by George Ziadeh.................................. 83

Part II: Selections from Aquinas Institute of Theology and
Holy Apostles College & Seminary

    Presentation 8:
    Social Media and Rural Evangelization: A Model for the
    New Evangelization by Fr. Dominic Ibok....................... 95

Presentation 9:
Online Communities: A Comparative Study of Catholic Communities on Facebook by Jeremy Chan..................119

Part III: Afterword

Presentation 10:
Love and Responsibility: The Personalization Principle in Cyberspace by Dr. Sebastian Mahfood, OP................... 143

Magisterial Resources 155

About the Editor.. 169

ABOUT THIS BOOK

This book was produced in part by the students of Saint Joseph's Seminary in Yonkers, NY, who participated in an intensive course on the theology of social media under Dr. Sebastian Mahfood, OP, in the fall of 2019. While almost all the students who participated in the course are Millennials (born between 1980 and 1996), some are GenZ (born 1997 and later, though different labels, like iGen, have been applied), and we're going to see a steady increase in this new generation, the oldest of whom just turned 23, in the coming years.

Seven of the 52 students who completed the course elected to publish their semester projects in this book, and their names are listed in the table of contents and in the chapters that follow. All are seminarians studying for the Catholic priesthood except for one lay woman, Joan Brisson, who is pursuing her M. A. in Theology. Of the seminarians, all but one traveled from another country to study in the United States, which testifies to the international character of this book with contributions from seminarians out of India, South Korea, Nigeria, Vietnam and Israel.

In addition, two contributions from Aquinas Institute of Theology and Holy Apostles College & Seminary are included to demonstrate the advancement of this teaching in the Doctor of Ministry and Master of Arts in Pastoral Studies programs in other theological schools. They were provided by a Nigerian priest and a Singaporean layman.

All contributors to this book, on completion of their degrees, are poised to make a difference in our world for the good of our faith, and one way in which they are likely to do so is through an effective use of social media as they describe in the pages below.

PREFACE

Sister Marysia Weber, RSM, DO, MA
Director of Consecrated Life
Archdiocese of St. Louis, MO

The Catholic Church embraces all that is good in new technologies and seeks to engage with persons using these technologies as a means of inviting them into a deeper relationship with Jesus Christ. More recently, this includes the Internet with its unprecedented opportunities to share information and build relationships beyond the confines of space and one's own culture. "Because of its capacity to surmount distances and put people in mutual contact, the Internet presents great possibilities...for the Church and Her mission...With the necessary discernment for its intelligent and prudent use, it is an instrument that can serve...for evangelization, missionary action, catechesis, educational projects and the management of institutes."[1]

The new technologies have also effected a profound cultural transformation not only in the means to communicate, but in how we relate to others through our devices. Consider the iPhone and its effect on our daily lives since its

---

[1] Pope Benedict, *The Priest and Pastoral Ministry in a Digital World: New Media at the Service of the Word*, 44th World Communications Day Message, 2010.

release in 2007. Many of us rarely go anywhere without our cell phones. A recent survey found that Americans check their phones 47 times a day and 50% of Americans check their phones in the middle the night. 18 to 24-year-olds check their phones 82 times a day and 75% of them check their phones in the middle of the night.[2] The average 8- to 10-year-old spends almost eight hours on various digital devices while teenagers spend eleven hours in front of screens.[3] This is more time than they do anything else, including sleep. We are noting effects in a younger demographic as well. Although the estimates vary, a recent study by the *American Academy of Pediatrics* cited that children begin interacting with digital media at four months of age. They also noted that 44% of children under age one use a mobile device on a daily basis to play games, watch videos or use apps. This percentage increases to 77% of two-year-olds.[4]

The Church is also attentive to the need for vigilance in

[2] Deloitte, 2016 Global Mobile Consumer Survey: US Edition; The market-creating power of mobile, 2016.

[3] Henry J. Kaiser Family Foundation, *Generation M2: Media in the Lives of 8- to 18-Year-Olds*, January 2010.

[4] Radesky JS, Silverstein M, Zuckerman B, Christakis DA, Infant self-regulation and early childhood media exposure Pediatrics. 2014 May;133(5):e1172-8. doi: 10.1542/peds.2013-2367. Reid Chassiakos Y, Radesky J, Christakis D, et al., AAP COUNCIL ON COMMUNICATIONS AND MEDIA. Children and Adolescents and Digital Media. Pediatrics. 2016;138(5): e20162593. Kabali HK, Irigoyen MM, Nunez-Davis R, et al. Exposure and use of mobile media devices by young children. Pediatrics. 2015;136(6):1044–1050. Gretchen Geng and Leigh Disney, "A Case Study: Exploring Video Deficit Effect in 2-Year-Old Children's Playing and Learning with an iPad," Proceedings of the 21st International Conference on Computers in Education 2013. http://espace.cdu.edu.au/ view/ cdu :40222.

using the technologies of the Internet. Borne of human ingenuity, media proposes and often imposes a mentality and a plan of life in constant contrast with the Gospel. Saint John Paul II warns, "The Internet offers extensive knowledge, but it does not teach values; and when values are disregarded, our very humanity is demeaned and man easily loses sight of his transcendent dignity."[5]

In likeness to the first disciples proclaiming our faith that Christ is the Savior of humanity, the Church seeks to avail itself with renewed creative commitment, turning to the Gospel and Christian tradition to guide, purify and elevate new forms of communication. This will require not only proper theological formation, but technological formation to be effective witnesses in digital environments. Decades before the advent of the Internet, the Church maintained an attentiveness to forming pastoral ministers, not only in the faith, but also in the use of media for evangelization. "If students for the priesthood and religious in training wish to be part of modern life and to be effective in their apostolate, they should know how the media work upon the fabric of society and the technique of their use. This knowledge should be an integral part of their ordinary education. Indeed, without this knowledge an effective apostolate is impossible in a society which is increasingly conditioned by the media."[6]

This mandate for training future priests in the use of media for effective evangelization is echoed throughout Church

---

5 John Paul II, "Internet: A New Forum for Proclaiming the Gospel," #4. (January 24, 2002). http://w2.vatican.va/content/john-paul-ii/en/messages/communications/documents/hf_jp-ii_mes_20020122_world-communications-day.html

6 Martin J. O'Connor, Pontifical Council of Social Communications, *Communio et Progressio*, #111. (May 23, 1971). http://www.vatican.va/roman_curia/pontifical_councils/pccs/documents/rc_pc_pccs_doc_23051971_communio_en.html

writings. "Who better than a priest, as a man of God, can develop and put into practice, by his competence in current digital technology, a pastoral outreach capable of making God concretely present in today's world and presenting the religious wisdom of the past as a treasure which can inspire our efforts to live in the present with dignity while building a better future?"[7]

Seminarians and priests today will necessarily employ digital social networks as a means of making the infinite richness of the Gospel reach the minds and hearts of all. In his 2020 World Communications Day Message, Pope Francis tells us that no one is exempt from making "communications an instrument with which to build bridges, to unite and to share the beauty of being brothers and sisters in a moment of history marked by discord and division."[8]

The collection of articles contained in *Among the Marvelous Things: The Media of Social Communications and the Next Generation of Pastoral Ministers* is a fruitful example of Saint Joseph's Seminary in Yonkers, New York, Aquinas Institute of Theology in St. Louis, Missouri, and Holy Apostles College & Seminary in Cromwell, Connecticut, taking seriously the Church's mandate to decisively insert into pastoral formation programs not only solid theological formation, but also formation opportunities for seminarians and lay ministers to incorporate digital technology into their work of

---

[7] Pope Benedict, *"The Priest and Pastoral Ministry in a Digital World: New Media at the Service of the Word."* (May 16, 2010). http://w2.vatican.va/content/benedict-xvi/en/messages/communications/documents/hf_ben-xvi_mes_20100124_44th-world-communications-day.html

[8] "Theme of the Holy Father Francis's Message for the 54th World Communications Day," (September 28, 2019). https://press.vatican.va/content/salastampa/en/bollettino/pubblico/2019/09/28/190928e.html

bringing the Light of Christ to the millions of persons across the globe.

May the Church's mandate to form the next generation of pastoral ministers in the use of current communication media for evangelization become more widespread.

Sister Marysia Weber, RSM, DO, MA is the author of Screen Addiction: Why You Can't Put That Phone Down. *(En Route Books and Media, 2019). Her work in the area of the media of social communications may be found on her author page at http://enroutebooksandmedia.com/screenaddiction/*

FOREWORD

Attachment, Receptivity, and Communion: An Exploration in Communication to Best Evangelize Through Internet Based Social Media Platforms

Dr. Timothy G. Lock, TFI
Director of Psychological Services
St. Joseph's Seminary - Dunwoodie

There is tremendous confusion – the wreckage can only be described as simply tragic. It is 1950, and children are dying. Literally (see Karen, 1998). From the London bombings of World War II, many children were orphaned, and the city sought to provide for their needs. Water, food and shelter were allocated and amply. But the children were still dying. Why this was happening was greatly perplexing.

What should be done? How do the otherwise traumatized survivors identify and fix the origin, or the cause, of the problem?

Fast-forward 70 years.

There is tremendous confusion – the wreckage can only be described as simply tragic. It is 2020, and people are dying. Not literally like in World War II, but today the assaults from the dictatorship of relativism are causing spiritual deaths at an alarming rate. Water, food and shelter are metaphorically provided by the Catholic Church. But the children are still dying. Why this is happening is greatly perplexing.

What should be done? How do the otherwise traumatized survivors identify and fix the origin, or the cause, of the problem?

In 1950, once the problem was identified, the British government reached out to the World Health Organization to help unravel the mystery and identify the cause which seemed to be related to the war (Bowlby, 1988). Babies and young children were literally dying. Not from disease. Not from illness. Not from anything related to the direct impact of the bombings, for these children had not been physically harmed in the bombings. Some were not even in London at the time of the bombings. These children were, we could say, secondary victims in that their parents were killed, and the children were left alone. While the British government exerted a significant effort to provide the necessary care for these orphans, many were still dying. And no medical cause could be identified.

Dr. John Bowlby, a new psychiatrist who had just begun working for the World Health Organization (WHO), was asked to investigate the matter and report back to WHO. Bowlby's results shocked the world. The primary negative impact on the kids was not just the loss of the caregiver but the loss of the care offered by the caregiver. The children who died were the ones who had not formed a relationship of any kind with the staff at the orphanages. You see, the British people were providing food and shelter and clothing to the new orphans, but due to their overwhelmingly large numbers, the staff were unable to provide the children with the personal touch (literally and figuratively), that is, with relationship, with love.

History does not fault the English as they were inundated with orphans and did everything they could to provide for them. Bowlby's observations and research shed a new light on the issue and found a fourth need without which resulted in child death. Soon, Bowlby and his colleagues' newly developed "attachment theory" regarding the newly identified "need" of the relationship between the parent and the child resonated with scholarship throughout the world. The problem in Great Britain was addressed and a new field of study was identified (that continues to flourish to this day). It took a group of brilliant minds willing to think outside the box to solve the riddle (see Bowlby, 1988).

From a Catholic perspective, this research makes sense. We are made for communion. Spiritually, we are created in the likeness and image of God: Father, Son and Holy Spirit. However, the three Persons of the Blessed Trinity are not stagnant, neutral and self-governed. The Most Holy Trinity reflects a dynamic and constant exchange in a relationship of love. As we—human people—are created in the likeness and image of the Trinity, so are we created for communion. Bowlby's research highlights this spiritual fact in a psychological way and offers psychological evidence for this spiritual fact.

In 2020, we are faced with a different problem and the problem has been known for some time. The Catholic Church has exerted a significant effort to provide the necessary care for the people of God, yet many are spiritually dying, and the cause is unclear. So, the Catholic Church, in Her wisdom, has sounded a clarion call asking her body to reflect on issues of missionary zeal, sharing the faith and evangelization.

Another way to reflect this truth of communion is to say that we are created for communication with one another, communication that fosters this communion. What happens if the communication becomes stagnant? What happens if the communication between individuals is somehow impeded?

The image and confusion of the Tower of Babel from Genesis 11 comes to mind. The world went from one common language to many languages, and the people could not understand one another.

With the introduction of the internet and the world wide web, a new "language" has developed. With this new language and new "place" to spend time, the Church has had to adapt to reach people who are no longer hearing Her message. To prevent this massive spiritual death, the Church as a body must spread the Good News of the Gospel; we must evangelize. However, to evangelize we must speak the same language. In 2020, this includes that new language of the internet and the world wide web.

And how does one learn this new language? Typically, learning takes place from someone who has mastered the language. Ideally, we want to learn this new language on the internet from people who have spent significant time in this digital land – with those who have "lived" on the internet. Once we have begun to speak the new language, we can reach out in a way that can be understood.

Another concept is associated with this need to engage people in a new way. St. Thomas Aquinas put it in this way, "Quidquid recipitur ad modum recipientis recipitur" (*Summa Theologiae*, 1273/1981, I-II, 75.a5), or in English, "What is received is received in the mode of the receiver." This relates to our current discussion because whatever we communicate will not be received unless it can somehow fit into the mindset of the receiver. The receiver must have a schema or a framework to understand what is being said. Our speaking the receiver's language allows us to more aptly present the Gospel in a way that can resonate with their mindset.

What was true in 1950 remains true in 2020: it will take a group of brilliant minds willing to think outside the box to solve the riddle. In August 2019, there was a sort of "meeting of the minds" with a group of students enrolled in a course in communications offered at St. Joseph's Seminary-Dunwoodie in Yonkers, NY. I was privileged to audit the course and, although I was not "in the game," so to speak, I was "on the sidelines" hearing the seminarians discuss their assimilation of the course lecture materials, assignments, readings, and their lived experience.

Right now, as you flip the pages of this book (physically or electronically), you are about to learn something of this new language from these seminarians who speak this language well. The current collection of essays offers you a unique portal into minds of those who live in this "internet world" – yet these are not casualties of the war. They may have earned a Purple Heart (or two) along the way, but they have come out the other side still fighting. They are soldiers and, more accurately, they are in officers-in-training.

As you read about this new language of the internet, you will learn about how these seminarians have used this language to advance the Gospel. You will be amazed at how they have integrated the Faith and shared the Faith online. You will marvel at the wisdom – wisdom beyond their years – which will refine how you to engage the culture.

Yes, in 2020, there is tremendous confusion – the wreckage can only be described as simply tragic. Dr. Sebastian Mahfood, OP, and the priests, seminarians and lay ministers of St. Joseph's Seminary – Dunwoodie, Aquinas Institute of Theology and Holy Apostles College & Seminary will accompany you as you seek to find a way to through this confusion. They will demonstrate for you approaches to evangelize that can help you communicate in such a way that the receiver receives what is being offered; the seminarians will help you articulate the Gospel so that it fits into the mindset of those who are the receivers.

This is an awesome task and an awesome responsibility. When entering into this new approach to communication, we must stay close to Our Lord, we must stay close to Our Lady, we must stay close to our Guardian Angel and, in the words of Pope St. John Paul II, we must, "Be not afraid!"

Reference

Aquinas, T. (1981). <u>Summa Theologica</u> (English Dominican Province, Trans.). Westminster, MD: Christian Classics. (Original work composed 1273)

Bowlby, J. (1988). <u>A Secure Base: Parent-Child Attachment and Healthy Human Development</u>. New York: BasicBooks.

Karen, R. (1998). <u>Becoming Attached: First Relationships and How They Shape Our Capacity to Love</u>. New York: Oxford University Press.

INTRODUCTION

Dr. Sebastian Mahfood, OP
Professor of Interdisciplinary Studies
Holy Apostles College & Seminary, Cromwell, CT

The title of this book is taken from the Second Vatican Council's decree on the media of social communications, *Inter Mirifica*, promulgated by Saint Paul VI on December 4, 1963. Within this document, the Council provided its rationale for engaging the topic as follows:

> It is . . . an inherent right of the Church to have at its disposal and to employ any of these media insofar as they are necessary or useful for the instruction of Christians and all its efforts for the welfare of souls. It is the duty of Pastors to instruct and guide the faithful so that they, with the help of these same media, may further the salvation and perfection of themselves and of the entire human family.[9]

---

[9] Paul VI, *Inter Mirifica*, §3 (December 4, 1963). Online at https://www.vatican.va/archive/hist_councils/ii_vatican_council /documents/vat-ii_decree_19631204_inter-mirifica_en.html

The media so named are "the press, movies, radio, television and the like," which remain important means for evangelization today. The difference between 1963 and 2020 is one of access to the means of production. In 1963, almost everyone was a consumer of these media, but it was expensive to be a producer of them as concerned the human, financial and physical resources necessary to produce quality programming. In 2020, anyone with an Internet connection can be a producer of these media as evidenced by the rapid development of social media commons such as personal websites and blogs, podcasting services, and interactive engagement platforms such as Instagram, Facebook, Twitter, YouTube, and the like.[10]

Important in the mandate provided by *Inter Mirifica* is the statement "[i]t is the duty of Pastors to instruct and guide the faithful...," which calls to mind the legal rule, popularized by St. Thomas Aquinas, "nemo dat quod non habet," which means "no one gives what he does not have." So, how do Pastors go about getting the knowledge and experience to instruct and guide the faithful in the use of the media of social communications? They must either bring it into their ministries with them or learn it along the way, and this need to gain such knowledge and experience reaches into the heart of the pastoral formation programs. If these programs are going to

---

[10] With very little initial expense, for example, I set up a Catholic publishing house called En Route Books and Media, available online at https://www.enroutebooksandmedia.com, which has at the time of this writing published over 120 Catholic books and is serving as the publishing house for this book, and a Catholic radio station called WCAT Radio (WCAT is short for Why CATholic?), available online at https://www.wcatradio.com, which currently hosts almost six dozen Catholic programs with over, at the time of this writing, 5,000 podcasts on its website available now for public consumption.

be effective in training ministers, then the first to be trained are the formators themselves.[11]

The key really is not so much in training pastoral ministers on the use of a specific tool – that is, on its technical aspects – but on the formation provided in how the media may be used to evangelize within a media age. Just like the apostle Paul used the media of his day to spread the Gospel message, writing instructional and motivational epistles to the Christian communities he formed, and even one to a community in Rome he didn't form, today's pastoral ministers are called to use the media of our day to engage their communities. In fact, Saint John Paul II wrote in *Redemptoris Missio* (1991) the following:

> The first Areopagus of the modern age is the world of communications, which is unifying humanity and turning it into what is known as a "global village." The means of social communication have become so important as to be for many the chief means of information and education, of guidance and inspiration in their behavior as individuals, families and within society at large. In particular, the younger generation is growing up in a world conditioned by the mass media. To some degree perhaps this Areopagus has been neglected. Generally, preference has been given to other means of preaching the Gospel and of Christian education, while the mass media are left to the

---

[11] The Catholic Distance Learning Network that was founded by me and Br. Bernie Stratman, SM, through the Seminary Department of the National Catholic Educational Association in 2006 in response to a suggestion within Saint John Paul II's 2005 apostolic exhortation "The Rapid Development," for example, trained between 2007 and 2014 over a hundred seminary faculty in online teaching and learning tools and methods.

initiative of individuals or small groups and enter into pastoral planning only in a secondary way. Involvement in the mass media, however, is not meant merely to strengthen the preaching of the Gospel. There is a deeper reality involved here: *since the very evangelization of modern culture depends to a great extent on the influence of the media, it is not enough to use the media simply to spread the Christian message and the Church's authentic teaching. It is also necessary to integrate that message into the "new culture" created by modern communications.* This is a complex issue, since the "new culture" originates not just from whatever content is eventually expressed, but from the very fact that there exist new ways of communicating, with new languages, new techniques and a new psychology. Pope Paul VI said that "the split between the Gospel and culture is undoubtedly the tragedy of our time," and the field of communications fully confirms this judgment.[12] (emphasis mine)

This necessity of integrating the message into the new culture created by modern communications was confirmed in "The Rapid Development" (2005), when Saint John Paul II wrote,

Such is the importance of the mass media that fifteen years ago I considered it inopportune to leave their use completely up to the initiatives of individuals or small groups, *and suggested that they be decisively inserted*

---

[12] John Paul II, *Redemptoris Missio: On the Permanent Validity of the Church's Missionary Mandate* (December 7, 1990). Online at http://www.vatican.va/content/john-paul-ii/en/encyclicals/documents/hf_jp-ii_enc_07121990_redemptoris-missio.html

into pastoral programs.[13] (emphasis mine)

The significance of Saint John Paul II's reaffirmation of this 'suggestion' is that it is his final instruction on the subject. He died just a few months later, leaving behind a long legacy as the communications pope. As of this writing in January 2020, in fact, "The Rapid Development" remains the most recent apostolic exhortation to the Catholic community to undertake a specific activity in regard to the use of communications media – and that can be summed up in two words: "own it."[14]

On the same day Saint John Paul II promulgated "The Rapid Development," he also promulgated the 39th World Communications Day Message in which he expressed an appeal that "the men and women of the media will play their part in *breaking down the dividing walls of hostility in our world*" and "use the resources at their disposal to strengthen the bonds of friendship and love that clearly signal the onset of the Kingdom of God here on earth."[15] (emphasis his) Those resources are, as Pope Pius XII explained in *Miranda Prorsus*

---

[13] John Paul II, "The Rapid Development" (January 24, 2005). https://w2.vatican.va/content/john-paul-ii/en/apost_letters/2005/documents/hf_jp-ii_apl_20050124_il-rapido-sviluppo.html

[14] This is not to discount, of course, the annual World Communications Day messages that have been promulgated every January 24, the feast of St. Francis de Sales, the patron saint of journalists, since 1967, letters that continue to provide direct instruction from the popes in the use of social media.

[15] John Paul II, "The Communications Media: At the Service of Understanding Among Peoples" (January 24, 2005). https://w2.vatican.va/content/john-paul-ii/en/messages/communications/documents/hf_jp-ii_mes_20050124_world-communications-day.html

(1957) "gifts from God"[16] and ought to be used as such.

Saint John Paul II had written in his 2002 World Communications Day message a most profound mandate based on an obvious rationale:

> The Internet causes billions of images to appear on millions of computer monitors around the planet. From this galaxy of sight and sound will the face of Christ emerge and the voice of Christ be heard? For it is only when his face is seen and his voice heard that the world will know the glad tidings of our redemption. This is the purpose of evangelization. And this is what will make the Internet a genuinely human space, for if there is no room for Christ, there is no room for man. Therefore, on this World Communications Day, *I dare to summon the whole Church bravely to cross this new threshold, to put out into the deep of the Net, so that now as in the past the great engagement of the Gospel and culture may show to the world "the glory of God on the face of Christ" (2 Cor 4:6).* May the Lord bless all those who work for this aim.[17] (emphasis mine)

To assist in this mandate, the Catholic faithful have been well-instructed by the Church. Consider the Pontifical Council of Social Communication's pastoral instruction entitled *Aetatis Novae*, which, though predating the advent of the world wide web by a few years, provides in its appendix a useful outline

[16] Pope Pius XII, *Miranda Prorsus* (September 8, 1957). http://www.vatican.va/content/pius-xii/en/encyclicals/documents/hf_p-xii_enc_08091957_miranda-prorsus.html

[17] John Paul II, "Internet: A New Forum for Proclaiming the Gospel" (January 24, 2002). http://w2.vatican.va/content/john-paul-ii/en/messages/communications/documents/hf_jp-ii_mes_20020122_world-communications-day.html

for designing pastoral plans for social communications.[18] The mandate for the plan designed in *Aetatis Novae* lay in the pastoral instruction on the means of social communication that was promulgated in a document entitled *Communio et Progressio,*[19] released through the Pontifical Council of Social Communications in 1971. The instruction leads with the emphasis the Second Vatican Council placed on social communication in explaining,

> A deeper and more penetrating understanding of social communication and of the contribution which the media it uses can make to modern society, can be derived from a number of documents issued by the Second Vatican Council. These are, notably the Constitution on "The Church in the World Today", 2 the Decree on "Ecumenism", 3 the Declaration on "Religious Freedom", 4 the Decree on "The Missionary Activity of the Church", 5 and the Decree on "The Pastoral Duties of Bishops". 6 And, of course, there is a Decree that is wholly devoted to a discussion of "The Media of Social Communication".

The promulgation of *Inter Mirifica* on December 4, 1963, then, was foundational on the importance of media of social communications. It set the tone for everything that would follow within these other documents and was placed in context with the fullness of the Church's mission by being

---

[18] John Foley, *Aetatis Novae* (February 22, 1992). Online at http://www.vatican.va/roman_curia/pontifical_councils/pccs/documents/rc_pc_pccs_doc_22021992_aetatis en.html

[19] Martin J. O'Connor, Pontifical Council of Social Communications, *Communio et Progressio* (May 23, 1971). http://www.vatican.va/roman_curia/pontifical_councils/pccs/documents/rc_pc_pccs_doc_23051971_communio_en.html

promulgated on the same date as *Sacrosanctum Concilium*, the constitution on the sacred liturgy. In a single stroke, then, the Church addressed the two greatest commandments identified by Christ when he affirmed, as reported in Luke 10:27, that we "shall love the Lord, [our] God, with all [our] heart, with all [our] being, with all [our] strength, and with all [our] mind, and [our] neighbor as [ourselves]" [NIV] by instructing us in new ways how to talk to God and man.

When Saint John Paul II 'suggested' in *Redemptoris Missio* (1990) that instruction in the use of mass media be decisively inserted into programs of pastoral formation, he was affirming what was by that point several decades of direct communications leadership on the part of the Catholic Church. The Second Vatican Council's *Inter Mirifica* (1963) provided the mandate to establish the Pontifical Council for Social Communications (now called the Dicastery for Communication) and the context for all the World Communications Day messages that followed from 1967 to the present year.

As with any teaching that requires an active response, what remains for pastoral ministers is to develop a plan and make it happen. This book, written by the next generation of pastoral ministers currently involved in and also preparing for pastoral leadership, opens a window into the future use of social media for the purpose of evangelization.

PART I

ST. JOSEPH'S SEMINARY – DUNWOODIE

YONKERS, NEW YORK

SEVEN SELECTIONS

1

Burning Questions of Teens Today, or "What do Catholic Teens Really Want to Know about Their Faith and How Do We Tell Them?"

Joan Brisson, MA in Theology, Archdiocese of NY

In the late 1980s singer/song writer Billy Joel composed a song entitled: *We Didn't Start the Fire*. The song begins "We didn't start the fire; it was always burning since the world's been turning."[20] This song is a reminder that all the good and bad that has happened over a half century, from the H bomb to the Bay of Pigs invasion to the Cola Wars seems crazy, but things have always been crazy in the human experience and things most likely will continue to be crazy as long as humans exist and the world turns. We could expand these verses to include the beginning of man's time when God put Adam and Eve in the Garden of Eden to the 21st century and the new world of Social Media. If this song continued through the 1980s until today, it would have some catching up to do. With the advent of the internet, AOL, Google and the explosion of

---

[20] Joel, Billy. "We Didn't Start the Fire." In *Storm Front*. Columbia Records, 1989.

social media platforms, humanity has seen many developments in the area of technology that have taken us to a new realm. These rapid developments have not been completely unpredicted by the Catholic Church, and several encyclicals speak to "fires" that have developed long before we knew what a post or a tweet was. These encyclicals are meant as a guide as we face our "fires," and especially as the youth deal with growing up in a world where both their offline and online lives are synonymous. Their "fires" are ever present. It is our youth who need guidance and accurate information as they live in this digital age. But to communicate with a generation of digital kids, one needs to communicate through the digital media to reach them. Pre-teens and teens have screens and media present in their lives constantly.

How do we best reach them and communicate important messages about our faith? We often hear the phrase "meet them where they are." Well, our pre-teens and teens are online. And they are online every day, all day. Most of them will only have an 8th grade education in their faith at best. Their important questions will be answered by peers who are equally ill equipped with accurate knowledge of the faith. Social media can be a tool to gather them in. To make them see that their "fires" are part of a long line of "fires" that have been burning since the world's been turning. They need real faith-based answers to questions that are important to them.

To get accurate Catholic teaching out to teens, there are several social media platforms, an example of which is Instagram, which would be most effective. I have created a sample post below for Instagram that contains a contemporary relevant question that teens may never feel comfortable asking an adult. The goal is to begin to get correct Catholic teaching out to Pre- and Post-Confirmation teens who may have unanswered questions about real topics/issues that occupy their thoughts. This is a sample that could be posted on a Curious Catholic Teen Instagram Account and could deliver correct Catholic guidance and moral answers.

Is it OK to get a tattoo if I'm Catholic?

In principle, the Catholic Church does not oppose tattoos.
But ....if you are thinking about getting a tattoo, consider the following:

The images should be moral and not evil

Be prudent dude! this is gonna be forever, make sure you are sure   Do your reserach !

Think !! Just because you can, should you?

Advice – It's OK to wait

And remember – it hurts!

Spelling and grammar aside, the message provides an answer to a question a teen is likely to ask, and its presence in social media is intentional. An intentional communication is a human act, and human acts are moral acts.

World Communications Day was born

In 1963, the Second Vatican Council issued its decree *Inter Mirifica* (Decree on the Media of Social Communication), which established the Church's annual tradition of celebrating communications media. The Vatican Secretariat of Communications expressed this theme as "an invitation to tell the history of the world and the stories of men and women. In accordance with the logic of the 'good news' that reminds us that God never ceases to be a Father in any situation or with regard to any man. Let us learn to communicate trust and hope for history."[21] Echoing this message are the opening thoughts of Pope Francis's speech at the at the 53rd World Communication Day in January 2019: "Ever since the internet first became available, the Church has always sought to promote its use in the service of the encounter between persons, and the solidarity among all."[22]

That the Church has always been in favor of and supportive of the use of media is evidenced in its documents from the earliest writings regarding media – then radio, television and cinema – to the fast paced, interactive technology of today. The Church has recognized the importance and influence this art has when connecting to the human experience. The Church encourages and supports the arts and media and the goodness brought forth to humanity through them yet cannot permit them to be a barrier to God and salvation of the human soul. "From the time when these arts first came into use, the Church welcomed them, not only with great joy, but also with a motherly care and watchfulness, having in mind to protect her children from every danger, as they set out on this new path of progress."[23]

---

[21] "World Communications Day: History", accessed December 5, 2019, https://wcdnyc.org/history/.

[22] Francis. Message of the Holy Father Francis for the 53rd World Day of Social Communications (January 24, 2019), accessed August 23, 2019, https://press.vatican.va/content/salastampa/en/bollettino/pubblico/2019/01/24/190124b.html.

[23] Pius XII. Encyclical Letter of His Holiness Pius XII: Introduction *Miranda Prorsus* (September 8, 1957), accessed December 1, 2019, http://w2.vatican.va/content/pius-xii/en/encyclicals/documents/hf_p-xii_enc_08091957_miranda-prorsus.html.

From the time of Pope Pius XII (1939-1958) to today, the Church instructs us to appreciate these new ways of communication, embrace them and treat them with a watchful eye. The Church knows that with this progress will come both good and bad. In the late 1950s, a cautious approach was called for. Calling for prudence on the parts of adults, clergy and teachers, the Church sent a message that although these new media were extremely entertaining, responsibility was necessary. Our free will, after all, needs the guidance of a conscience formed in the mind of the Church. As these "fires" of the late 1950s raged, families could watch the nightly news in their living rooms, read the daily paper and see world news reels in movie houses. How would this influence us? How would we let this guide our paths? It became more important for us to rely on our moral theology for guidance.

As new forms of media rapidly developed, we had to more carefully choose with the guidance of our pastors what entered our homes and take responsibility as friends, parents and families. Today, that work continues as we discern how to turn away from that which drains our time, worth and soul, especially from that which leads us to sin whether by commission or omission. Our spending time with video games or in front of the TV needs to be balanced with time spent at the dinner table or in taking walks together. It is the peoples' responsibility to understand what is at the heart of the communication—whether it is a movie, TV show, podcast, YouTube video—and to ask what is their purpose and what is their intent? To spread good or evil? As media develops, so does our need to be more diligent.

The evil that came with radio or television is greatly outpaced by the internet and our tuning out no longer guarantees our ability to step away from the problem. Taking a walk with the family today almost always guarantees that a smartphone is brought along. Stepping back is harder than ever. Parameters for use and precautions must be put in place as with the use of all media. It is essential for adults and parents to give clear examples when using any media. Allowing your six-year-old to be in the room while an R rated movie is on television desensitizes him or her to the evil (sex, violence, language) that is all too commonplace.

In May of 1967, on the first World Communications Day, Pope Paul VI called for workers in the field of communication, namely teachers and parents who are "the first, irreplaceable educators of their children"[24] to exemplify how to use "these means sensibly with moderation and self-discipline."[25] The importance of right guidance with media and social media is more critical than ever before. "Today's media environment is so pervasive as to be indistinguishable from the sphere of everyday life. The Net is a resource of our time."[26]

We use social media in every aspect of life. It is a resource for our health, love, dating, education, socializing and faith. Before we can go forth and live and teach within this mediated world, we must ourselves understand that the Church is supportive of media. We must know what our own practices are and should be and where we fall short. And we must be good examples to our children through proper use of and

---

[24] Paul IV. Message of the Holy Father for the World Social Communication Day: Theme: Church and Social Communication: First World Communication Day (May 16, 1974), accessed November 30, 2019, https://www.vatican.va/content/paul-vi/en/messages/communications/documents/hf_p-vi_mes_19670507_i-com-day.html.

[25] Ibid.

[26] Francis. Message of the Holy Father Francis for the 53rd World Day of Social Communications.

instruction in all media to understand its values and dangers.

The wonderful aspects of social media are seen in every story of reunited families or soldiers on FaceTime from across the miles. It allows for self-expression through art and poetry. It can make people feel less lonely in finding a group to join and play a game or share knitting tips. The internet can provide easy access to vital information on health. Couples can meet and fall in love on sites such as Catholicmatch.com. Communities can come together to support each other across communities or across the globe. Teens can find connections with the world. In 2015-2016, the Holy Year of Mercy saw Pope Francis calling for acts of compassion, tenderness and forgiveness. "Communication has the power to build bridges, to enable encounter and inclusion, and thus to enrich society. How beautiful it is when people select their words and actions with care, in the effort to avoid misunderstanding, to the wounded memories and to build peace and harmony." [27]

Pope Francis highlights the good but knows that all too easily our technological communication can be used in anger to hurt or gain vengeance. Our Facebook feeds are also cluttered with stories of "cyberbullying," which is a term developed from the playground bully of old. No longer is bullying left on the playground; rather, it travels around with us, comes home with us and destroys us. Cyberbullying knows no age limit. A wrong political opinion can have one ousted from a group or kicked out of a social circle. Isolation can lead to pursuing the wrong avenues of entertainment. When Pope

---

[27] Francis. Message of the Holy Father Francis for the 50th World Day of Social Communications: Communication and Mercy: A Fruitful Encounter (January 24, 2016), accessed December 1, 2019. http://www.vatican.va/content/francesco/en/messages/communications/documents/papa-francesco_20160124_messaggio-comunicazioni-sociali.html.

Pius XII wrote about the responsibility of film makers, directors, producers, as well as the viewers, he did not foresee the explosion in the world of pornography. Or when singling out clergy to help in guiding our choices in the media, he could never had anticipated the worldwide scandal the Church itself would face dealing with sexual abuse. "Unless the mounting development of technical skill, applied to the diffusion of pictures, sounds and ideas, is subject to the sweet yoke of the law of Christ, it can be the source of countless evils."[28]

For all the good and bad our technological age brings, we find ourselves, and more importantly our children, irreversibly enmeshed in it. We must include God in this every expanding horizon. There is a place for technology within faith. Generations who come after us will live and love and prosper within the world where technology and social media is ever present. Our faith must be ever present, too. "Faith itself, in fact, is a relationship, an encounter; and under the impetus of God's love, we can communicate, welcome and understand the gift of the other and respond to it."[29]

How do we teach our youth to have a relationship with God and grow in the knowledge of their faith? If we are to "meet them where they are," we must answer their questions about life, love and the world in which they live through avenues with which they are familiar. We must be where they are. Teens today spend many of their waking hours looking at some type of screen. Media are used in classrooms beginning in early education and throughout all levels of education.

---

[28] Pius XII, *Miranda Prorsus*. (September 8, 1957). http://www.vatican.va/content/pius-xii/en/encyclicals/documents/hf_p-xii_enc_08091957_miranda-prorsus.html

[29] Francis, Message of the Holy Father Francis for the 53rd World Day of Social Communications. http://www.vatican.va/content/francesco/en/messages/communications/documents/papa-francesco_20190124_messaggio-comunicazioni-sociali.html

Children are taught how to function using technology. Classes may take place online or through video link. Homework often includes watching videos. Social interaction is over texting or messaging or through apps like Discord. Entertainment may include interactive video games with friends who are across town or across the country and the solitary streaming of the latest episode of *Riverdale* on Netflix.

According to a Pew research study, teens spend an average of 3 hours a day online:

> "Teens now enjoy more than five and a half hours of leisure a day (5 hours, 44 minutes). The biggest chunk of teens' daily leisure time is spent on screens: 3 hours and 4 minutes on average. This figure, which can include time spent gaming, surfing the web, watching videos and watching TV, has held steady over the past decade. On weekends, screen time increases to almost four hours a day (3 hours, 53 minutes), and on weekdays teens are spending 2 hours and 44 minutes on screens."[30]

Whether or not we agree that this is too much time or just a fact of life, this is their reality. The fact is teens are on their screens for much of their waking day and this is how information comes to them. For those of us who grew up with one screen, this is a whole new world. If the one screen you had was, like mine, in your living room and

---

30 Gretchen Livingston, "Pew Research Center, Fact Tank - News in the Numbers: *The Way U.S. teens spend their time is changing, but differences between boys and girls persist,*" (February 20, 2019), accessed December 5, 2019, https://www.pewresearch.org/fact-tank/2019/02/20/the-way-u-s-teens-spend-their-time-is-changing-but-differences-between-boys-and-girls-persist/.

your viewing day ended with the national anthem and an off the air signal, the world of 24/7, 365 and more channels than you could possibly watch is daunting. Not only keeping pace with this but making a difference through it seems impossible. How can a non-Gen X, Y (Millennials) or Z successfully communicate and evangelize within his or her social media driven world? By fighting "fire" with "fire". We need to remember that each generation has its own unique growth and be open to using new apps and technology. We need to get educated on the social media platforms that are out there and how they work. Ask teens what they are using. Some will tell you, some won't. If the goal is to help teens strengthen and grow their faith and relationship with God, we need to help them live a faithful life through strengthening knowledge of their own faith through means they are comfortable with (even if we are not).

> "According to the right of information which every person has, communication must always respond in its content to the truth, and out of respect for the justice and charity it ought to be complete. With all the greater reason, this holds true when the communication is addressed to young people, who are in the position of opening themselves to the experiences of life." [31]

We can communicate faster and easier than ever before. In doing so, we must ensure that the information we use is

[31] John Paul II. Message of the Holy Father John Paul II for the 19th World Communications Day: Social Communications for a Christian Promotion of Youth (May 19, 1985), accessed December 2, 2019, https://w2.vatican.va/content/john-paul-ii/en/messages/communications/documents/hf_jp-ii_mes_15041985_world-communications-day.html.

theologically correct and true and age relevant. Teens may not be ready to receive a message we wish to give. Information must first be important to them. We need to know our audience and be truthful and true to God's word in our communication and posts.

> "Your media have to be imbued in orthodox content, period. Everyone is called to preach the gospel, but we should take extra precaution to not 'lean on our own understanding' (Prov. 3:5), and, just the same, we should steer clear of communicating novelties to others. Theological, apologetic, and evangelistic communication and experimentation should never be done at the expense of truth."[32]

If we are not true in our communication, that will come through. Teens are very perceptive and very advanced in many ways. They have their own "fires" they are dealing with. Depression, anxiety, high suicide rates, global warming, political divisions, gender confusion – are just a few issues facing our teens daily. The pressure to get into college and have pithy funny quips that get lots of "likes" is ever present. Having a post-worthy life seems to be the goal. Everyone always looks very happy in their posts – everyone seems to be living their best lives. This world they are living in is the future of our Church. More importantly, this is the now of our Church. As Pope Francis observed in his apostolic exhortation *Christus Vivit (Christ Is Alive)*, "We cannot just say the young people are the future of our world. They are its present; even

---

[32] Shaun McAfee. *Social Media Magisterium: A No-Nonsense Guide to the Proper use of Media* (St. Louis: En Route Books and Media, LLC, 2018), 121.

now, they are helping to enrich it."[33]

Our building Instagram accounts that are relevant to teens to educate and evangelize will get good Catholic content out to them. Curious Catholic teens need to be drawn in by posts that contain topics they want to know more about. By using hashtags, colorful attractive artistry and solid truthful theological information, I hope to build slowly a following that is shared throughout my local community and possibly grow from there. My daughter, who is an inspiration to me, is 15 years old. She has grown up with social media. She is a constant resource for me in understanding teens today. She herself has recently started an Instagram account where she posts daily about a character from a fiction series that she loves. Over a three-month period, she has grown her account to over 500 followers. And it is still growing. I am one of the lucky parents because she talks to me—admittedly, not about everything, but about many relevant and important issues. She wants to know her faith, and she wants to know the world and make good choices. I am encouraged that with all that bombard teens today they are not so completely tainted. They are there waiting for us to meet them online with God's words and true doctrine. The Billy Joel song ends with these words "when we are gone it will still burn on and on and on."[34] We must leave the truth of the Gospel to burn on for the next generation. We must leave it where they will find it.

---

[33] Francis. Post-Synodal Apostolic Exhortation of the Holy Father Francis to young people and the entire people of God, *Christus Vivit* (March 25, 2019), accessed November 30, 2019, http://www.vatican.va/content/francesco/en/apost_exhortations /documents/papa-francesco_esortazione-ap_20190325_christus-vivit.html.

[34] Joel, "We Didn't Start the Fire."

2

The Century of Technology and Communication

Tony George, Seminarian, Kerala, India,
Syro-Malabar Catholic Church

Introduction

We are living in the century of technology and communication. Social media have become a part of the basic need of society. It is difficult to think about a society without Facebook, YouTube, Instagram and other media of social communications. In this paper, I discuss the teachings of the Church concerning social communications. This paper is an answer for questions like the following: What is the attitude of the Church in the promotion of social media? How should the priest and pastor approach social media? What is the position toward youngsters who use social media? What is the role of social media in the work of evangelization?

This article also deals with the teachings of the Catholic Church that promote social communications. The Church believes that the use of social media is the best method of evangelization in the 21st century. This article also emphasizes the current state of the Church in the use of social media. It mentions some of the websites, blogs and other modes of social communications of the Catholic Church. The purpose of the article is to encourage readers to share the message of our loving Jesus through the new technology. Evangelization is the duty of every Catholic, and we must use technology to share the word of God.

Artifact and Description

https://www.instagram.com/jesus_chunks/

Jesus chunks is an Instagram page that advances the teachings of the Catholic Church through memes, images and small videos. This page shares the faith and love of Jesus through Instagram. The page will inspire you to love Christ and the bible and help the children to know the bible and saints through memes and short videos.

The Church on Evangelization through Media

"Go therefore and make disciples of all the nations, baptizing them in the name of the Father and the Son and of the Holy Spirit, teaching them to observe all that I have commanded you; I am with you always, to the close of the age." (Mathew 28: 19-20) These words of Jesus show the duty of the Church in the process of evangelization and in the sharing of the word of God. Evangelization starts with the proclamation of the Good News and continues through life according to the word of God. So, the first step of evangelization is communicating, preaching or announcing the word of God to those who are unfamiliar with sacred scripture and Jesus Christ. Pope Paul VI in his apostolic exhortation *Evangelii Nuntiandi* defines "evangelization in terms of proclaiming Christ to those who do not know Him, of preaching, of catechesis, of conferring Baptism and the other sacraments."[35]

The decree of the Vatican Council on the apostolate of the laity, *Apostolicam Actuositatem,* clarifies that "The apostolate of the Church and of all its members is primarily designed to manifest Christ's message by words and deeds and to communicate His grace to the world."[36] It shows that the Church and its members are responsible for communication of the word of God. Pope Paul VI in his apostolic exhortation *Evangelii Nuntiandi* explains "evangelizing means bringing the Good News into all the strata of humanity, and through its influence transforming humanity from within and making it new."[37] Communication is inevitable for evangelizing, transforming humanity and creating a world with love.

There should be a medium for communication, and the primary medium of communication is the body. Christian communication started with Jesus who used parables and sermons to convey the word of God. Communication is continued through the apostles in Christianity, and they speak to the people having different culture and language. Homilies from the pulpit used to be the most common way of evangelization. The rapid developments in communicative media and revolutions in technology offer more possibilities for

---

[35] Pope Paul VI, *Evangelii Nuntiandi:* Apostolic exhortation (December 1975). Accessed September 10, 2019, at *w2.vatican.va/content/paul-vi/en/apost_exhortations/ documents/hf_p-vi_exh_19751208_evangelii-nuntiandi.html.*

[36] *Apostolicam Actuositatem,* Decree of the Vatican Council on the apostolate of the laity (accessed September 10, 2019). *http://www.vatican.va/archive/hist_councils/ii_vatican_counc il/documents/vat-ii_decree_19651118_apostolicam- actuositatem_en.html*

[37] Pope Paul VI, *Evangelii Nuntiandi:* Apostolic exhortation (December 1975).

evangelization nowadays. Here we are going to deal with the teachings of the Church and make a detailed study concerning Church and evangelization through social media.

MIRANDA PRORSUS

Miranda Prorsus is an encyclical letter of his holiness Pius XII written in 1957 about Church teachings in motion pictures, radio and television. This document starts with a Catholic view about the technology: "Those very remarkable technical inventions which are the boast of the men of our generation, though they spring from human intelligence and industry, are nevertheless the gifts of God, Our Creator, from Whom all good gifts proceed."[38] Every technological development is a gift of God and should be used for the proclamation of the word of God.

The encyclical addresses the need and use of visualizing Catholic programming in the television. The pope says that "from religious ceremonies, as seen on Television, valuable fruits for the strengthening of the Faith and the renewal of fervour can be obtained by all those who, for some reason, are unable to be actually present; consequently, We are convinced that We may wholeheartedly commend programmes of this kind."[39] According to the encyclical, the priest should have knowledge about the modern technology and use the relevant technology to care for souls. It is the responsibility of the clergy to be aware about the modern technology when it is

---

[38]Pius XII, *Miranda Prorsus: Cinema, Radio and Television,* (September 1957), accessed September 11, 2019, at *w2.vatican.va/.../en/encyclicals/documents/hf_p-xii_enc_08091957_*miranda-prorsus.*html*

[39] Pius XII, *Miranda Prorsus: Cinema, Radio and Television* (September 1957).

closely related to our religious and social life. The pope makes clear the role of the clergy by this words, "The priest with 'the care of souls' can and must know what modern science, art and technique assert whenever they touch on the end of man and his moral and religious life."[40] This encyclical shows that the Church has a clear viewpoint about communication and different technologies of social media and how they should be used for evangelization.

INTER MIRIFICA

Inter Mirifica is a beautiful decree on the media of social communication, promulgated by his holiness Saint Paul VI on December 4, 1963. Paul VI explains how we can use these media to advance Christianity. The Church recognizes that "media, if properly utilized, can be of great service to mankind, since they greatly contribute to men's entertainment and instruction as well as to the spread and support of the Kingdom of God. The Church recognizes, too, that men can employ these media contrary to the plan of the Creator and to their own loss."[41] The Catholic Church has a vision of evangelizing through the media, so the Church supports the use of media for the proclamation of God's word.

This decree also speaks about the responsibility of the bishops, priests and believers in the subject of evangelization through social media. It says that the Church has "an inherent right ... to have at its disposal and to employ any of these

---

[40] Pius XII, *Miranda Prorsus: Cinema, Radio and Television,* (September 1957).

[41]Second Vatican Council. *Inter mirifica: Decree on the media of social communication,* (December 4, 1963), accessed September 13, 2019, *www.vatican.va/archive/hist_councils/ii_vatican_council/documents/vat-ii_decree_19631204_inter-mirifica_en.html*

media insofar as they are necessary or useful for the instruction of Christians and all its efforts for the welfare of souls. It is the duty of Pastors to instruct and guide the faithful so that they, with the help of these same media, may further the salvation and perfection of themselves and of the entire human family."[42] The decree declares that the media should give the right information and contribute more effectively to the common good.

The decree also explains the importance of the moral outlook and obligations that the media must follow. "Since the mounting controversies in this area frequently take their rise from false teachings about ethics and aesthetics, the Council proclaims that all must hold to the absolute primacy of the objective moral order."[43] According to the teachings of the Church, the youth have to discuss with their teachers and subject area experts what they see, hear or read for deeper understanding, and they have to be aware about the morality of society. For these reasons, the decree requires the Church to use social media for evangelization. "All the children of the Church should join, without delay and with the greatest effort," it exhorts, "in a common work to make effective use of the media of social communication in various apostolic endeavors, as circumstances and conditions demand."[44] The council says that the Church should arrange the program of social communication according to different cultural back-grounds and different age groups, and there should be a special consideration for the youth. Those in the Church must be encouraged in the designing of social media, for the sacred synod observed that social media must be used for evan-gelization, which is a responsibility of every member of the

---

[42] Second Vatican Council. *Inter mirifica.*
[43] Second Vatican Council. *Inter mirifica.*
[44] Second Vatican Council. *Inter mirifica.*

Catholic Church.

COMMUNIO ET PROGRESSIO

Communio et Progressio is a pastoral instruction written by order of the Second Vatican Council. The Council found the media to be a gift from God, which helps the brethren to proclaim the word of God. The Second Vatican Council had clarified how deep of a role media plays in making relationships, which *Communio et Progressio affirmed:* "Social communications tend to multiply contacts within society and to deepen social consciousness. As a result, the individual is bound more closely to his fellow men and can play his part in the unfolding of history as if led by the hand of God."[45] The Church teaches us that media of social communications will bring great unity to humanity.

The Second Vatican Council said that Christ revealed himself as the best example of communication without fear through his incarnation. The new pastoral instruction observed the nature of the Bible and the life of Christ: "During His life on earth, Christ showed himself to be the perfect Communicator,......It is now necessary that the same message be carried by the means of social communication that are available today. Indeed, it would be difficult to suggest that Christ's command was being obeyed unless all the opportunities offered by the modern media to extend to vast numbers of people the announcement of his Good News were

---

[45] Pontifical Council for Social Communications. "Communio et Progressio." (May 1971). Accessed September 14. 2019 www.vatican.va/roman_curia/pontifical_councils/pccs/documents/rc_pc_pccs_doc_23051971_communio_en.html

being used."[46] The instruction also declared that the seminaries and Catholic educational institutions should give priority to the most modern technology and should educate students about it.

The Church always expresses the great importance of the media. The apostolic letter of Saint John Paul II entitled *The Rapid Development* gives great information, for instance, about the development of social media and how the Church observes it. Likewise, Pope Benedict XVI talks about new technologies and the new relationships they enable in his message for the 43rd World Communications Day. It deals with how social media make a great change in society and how they encourage relationships, culture and friendships. Pope Benedict XVI encouraged people to use the media to be a witness of Christ.

AETATIS NOVAE

Aetatis Novae is a pastoral instruction of the Church on social communications that was promulgated on the 20th anniversary of *Communio et Progressio*. It gives accurate information about the Church mentality toward social communication and instructs that every episcopal conference and diocese must develop an integrated pastoral communications plan through consultation with the professionals in the communication field along with international and national Catholic organizations.

The Church always gives a prominent role to the media and media education. Based on the teachings of the Church, evangelization is the primary duty of every Christian, and to

---

[46] Pontifical Council for Social Communications. "Communio et Progressio." (May 1971).

proclaim the word of God we ought to use the most suitable methods of the day. In the early Christian community, the apostles used preaching as the way to proclaim the word of God, and St. Paul wrote letters; in this twenty-first century, we have a vast communications network at our disposal to proclaim the good news and the Church teaches us to use those resources to share that news.

Current State of Social Media Use

There are lot of websites and blogs that are promoting the Christian faith. Here I am just trying to describe the Catholic programs, blogs and websites that share Christ with the users of social media. Catholic Online is a beautiful website that gives a guide for confession, and it also deals with daily prayers and daily readings. This website can also be used as a Catholic encyclopedia. (https://www.catholic.org)

Christian Kids TV is an amazing YouTube channel that shows the history of saints through animations and cartoons and is really helpful for kids to know the lives of the saints so that they may imitate them. This channel motivates the children to do like the saints did in their lives. (https://www.youtube.com/channel/UC63iYSE884kZoq6rfy wsXDg)

Christian Cinema is a wonderful website for the youth to get knowledge about whether new movies have good content and gives a great opportunity to rent digital movies with good moral value. (https://www.christiancinema.com/)

Christianity Today is a beautiful blog that provides thoughtful biblical perspectives on the theology, culture and ministry of the Catholic Church. It also provides relevant news of the Church. (https://www.christianitytoday.com/ct/blogs)

These are some examples of how the Christian community uses social media for proclamation of the word of God. There are a lot of websites, Facebook pages, Instagram accounts, YouTube channels and blogs that are used by the Catholic community to share news about Jesus. We are living in a very modern society with the most modern technology. As Berrin Beasley and Mitchell R. Haney point out, "Nothing is more fundamental to an individual's physical well-being than life and having one's basic needs met."[47] Social media have become a fundamental need in our present society and have a great impact on how people connect with each other. Social media use images, text, emojis and memes to transmit meaning.

Social media can be defined as "web-based communication tools that enable people to interact with each other by both sharing and consuming information."[48] New media are used by the Church to connect with people and every church or parish has their own websites and blogs. We are living in a society with access to hundreds of channels in a home, and we can add to those channels. For that reason, today's media enable people to not only be consumers of content and services but also producers.

Parishes now maintain their own Facebook and Instagram pages to interact with people. Events and programs are shared through the pages. To reach the youth, parish webpages should be filled with attractive videos and memes conveying the love of Christ. If the charitable works of the Church are not reaching the youth, then Facebook and Instagram pages should be designed to promote the things that we are doing and include a short rationale from our Catholic social

---

[47] Berrin A. Beasley and Mitchell R. Haney. eds., *Social Media and Living Well (2015, USA)*, 49.

[48] Daniel Nations, "What is Social Media?" *Lifewire* (December 19, 2019). https://www.lifewire.com/what-is-social-media-explaining-the-big-trend-3486616

teaching.

A lot of institutions teach the Bible and theology, and online Bible classes exist in social media. "ComeuntoChrist," for instance, is a free online bible class for youngsters and is online at https://www.comeuntochrist.org/holy-bible. Many opportunities are given by the Church, but they are not always able to reach the youth, so we must motivate the youth to study the Bible as part of their social media experience. We should also make more opportunities within social media to teach Bible and create inspirational videos for the ease of promoting biblical studies.

The Church encourages people to hear the radio, so the Church owns different types of radio stations. The radio helps people to hear the word of God over the course of their daily activities and it motivates them to pray. The Church shares the word of God and biblical interpretation through radio. Relevant radio is a beautiful American Catholic radio station that broadcasts Catholic news, devotional songs and biblical interpretation. (https://relevantradio.com)

Today, the Church gives great importance and encourage-ment to those working in every field of social communication. The Church has its own websites, Facebook pages, Instagram accounts, movies, YouTube channels, blogs etc. Popes, bishops, priests, deacons and religious give a great deal of importance to the social media as the best way for evangeliza-tion. The Catholic Church encourages the youth to become the sound of the Church in the world of social media. Social media have become the best platforms for evangelization, so the Church believes that it is its duty to promote social communications and social networking.

Conclusion

The Church teaches us that we must use our modern

technology for the purpose of evangelization. Everybody is using social media in this 21st century, so there is a vast area of evangelization that can be done with memes, images and short videos through social media. It has long been proven that we can share the Gospel through images. Such images should help people to think about their prayer lives and faith. Social media should help encourage Catholics to believe in Christ and think about the saints.

An Instagram page will assist in disseminating images that are closely related to the Bible to help children grow in their faith. The memes on Instagram pages will make the viewers proud to be Catholic. It will also help to post such memes and images in parish bulletins and Facebook pages where the people come together. Prayers to the saints will also be on Instagram pages and they will encourage viewers to engage in prayer. New memes and images will be posted every day and new short videos that encourage the faith will be posted once every three days. Memes, images and videos will be closely related to the passages of the day, and they will help viewers come to know the Bible and read the Bible in their free time.

I hope through beautiful memes, images and short videos I can reach many people. I will add my email id with the memes and images to give feedback on them. I will collect the feedback on the page through the responses of viewers and other media. My aim is to encourage the study of Bible through memes, images and short videos and inspire the youth to share the word of God.

3

How to Get Catholic School Teachers to Subscribe to a Catholic Education Website

Steven McClernon, Seminarian,
Diocese of Rockville Center

Introduction

In the twenty-first century, public and private schools utilize educational websites to enhance their lesson plans and evaluate student progress. For example, the Connetquot Central School District, in Ronkonkoma, New York, subscribes to brainpopjr.com, abcya.com, and razkids.com so that teachers have fun and engaging resources to use during class and students can be enthusiastic about learning.[49] The concept of a Catholic educational website—one we have yet to build—is to give Catholic school teachers a fun and engaging resource for a classroom environment. The target audience to subscribe to this website would be Catholic elementary school teachers for grades one through six. The target audience to use

---

[49] "Student Resources," Connetquot Central School District, accessed August 23, 2019, http://ccsdli.org/student_resources/student_resources.

this website would be Catholic school children, ages seven to twelve. This website would be beneficial to schools, teachers, parents, and students, as this would be a valuable tool to reinforce the faith in a new and modern way. Within the first year of this website's launch, one outcome would be to have half of all Catholic elementary schools within a given diocese subscribe to this website and incorporate it into their religion curriculum.

Artifact and Description

https://www.youtube.com/watch?v=O4qbNmqT-s4

This is a link for a back-to-school message for teachers. It talks about the importance of incorporating one's Catholic faith into every aspect of teaching, especially as a Catholic school teacher.

Church Teaching of Evangelization in Social Media

In November of 2018, twenty-six-year-old missionary, John Chau, traveled to the North Sentinel Island to try and evangelize the people. In one of his diary entries he wrote, "God, I thank you for choosing me, before I was even yet formed in my mother's womb to be Your messenger of Your Good News."[50] Chau was trying to promote what was promulgated during the Second Vatican Council in *Inter Mirifica*, namely, "The Church recognizes that this media, if properly used, can be of great service to mankind...to spread and

---

[50] Doug Bock Clark, "The American Missionary and the Uncontacted Tribe," *GQ*, August 22, 2019, accessed August 23, 2019, https://www.gq.com/story/john-chau-missionary-and-uncontacted-tribe.

support the Kingdom of God."[51] According to the *Catechism of the Catholic Church,* "'Reborn as sons of God, [the baptized] must profess before men the faith they have received from God through the Church' and participate in the apostolic and missionary activity of the People of God."[52] Faith is meant to be a public, not a private act. This is best demonstrated during the Sunday Mass which is done in a community setting. As Aristotle states in the *Nicomachean Ethics,* one cannot be a virtuous man by himself, but only within a communal environment. Jesus Himself instructs His followers to "Go, therefore, and make disciples of all nations, baptizing them in the name of the Father, and of the Son, and of the holy Spirit, teaching them to observe all I have commanded you" (Matt 28:19-20 NAB). In a sense, the Church posits that every baptized person is a teacher of the faith. This is not a suggestion for us to go into the world and preach the faith, but it is a command that we do so.

The Catholic Church does not change, but it does find a method to adapt to the current age and use resources available to allow the proclamation of the Good News. The Church writes, "All children of the Church should join, without delay and with the greatest effort in a common work to make effective use of the media of social communication in various apostolic endeavors" (IM 13). There is an emphasis on using the most current means necessary to reach those who have

---

[51] Vatican Council II, Decree on the Media of Social Communications *Inter Mirifica* (December 4, 1963), 2, accessed August 23, 2019, http://www.vatican.va/archive/hist_councils/ii_vatican_council/documents/vat-ii_decree_19631204_inter-mirifica_en.html.

[52] *Catechism of the Catholic Church.* 2nd ed. (New York: United States Catholic Conference, 1997), 1270.

either fallen away from the faith or who do not yet know Jesus as their Lord and Savior. The twenty-first century world has multiple means to spread the salvific message of Jesus, including television, movies, radio, internet, social forums, websites, and social media platforms, including Facebook, Twitter, and Instagram. Technology is changing at a rapid pace, and, as a Church, we are called to stay engaged with the modern world, meeting people where they are, yet holding firm to the truths of our faith. The Church calls on people to not be fearful of modern technology, but to embrace it and use it as a tool to promote Jesus' message. We live in a world that is interconnected, which makes this time in history the perfect time to evangelize without traveling to different countries.

Social media is a great way to promote social justice causes. In *Aetatis Novae*, the Church instructs the faithful to use social media as a platform to promote good and important causes. It states, "Indeed, the power of the media extends to defining not only what people will think but even what they will think about...It is important therefore that Christians... give a voice to the voiceless."[53] The Church always promotes the dignity of all human life, and that includes in the use of social media. All people deserve to have respect and be treated properly. Furthermore, when there is injustice in the world, it is the Christian's job, using social media to bring attention to that need. Additionally, the Church writes, "Advertising...can contribute to the creation of new jobs, higher incomes and a more decent and humane way of life for all."[54] Furthermore,

---

[53] John Paul II, Pastoral Instruction on Social Communications on the Twentieth Anniversary of Communio Et Progressio, *Aetatis Novae* (February 22, 1992), 4, accessed August 23, 2019, http://www.vatican.va/roman_curia/pontifical_councils/pccs/do cuments/rc_pc_pccs_doc_22021992_aetatis_en.html.

[54] Pontifical Council for Social Communications, *Ethics in Advertising* (February 22, 1997), 5, accessed August 23, 2019,

the Church posits, "It is imperative that media respect and contribute to that integral development of the person" (AN 6). Promoting dignity for all life is one of the social cornerstones of the Catholic Church.

Now the Church also warns that this tool, social media, like any other tool, may be used for harm. One problem of media is the rise of cultural domination. The Church states, "Cultural domination is an especially serious problem when a dominant culture carries false values inimical to the true good of individuals and groups."[55] The Church wants all nations to share resources and ideas so that every human may benefit from this new technology. The Church is also against restrictions on the freedom of expression. "We deplore any attempts by public authorities to block access to information—on the Internet or in other media of social communication" (EI 12). Personal freedom is a God-given right, as is seen in Genesis. People have the freedom to choose God's love, otherwise, if love were forced onto humans, it would not be free, and thus, genuine. Similarly, the freedom of expression can be stifled by governments, as it has been in the past, for example, in China during the May uprisings of 1989 or during 1930's Nazi Germany. Stifling these freedoms is detrimental to a society, and the Church is trying to protect personal liberties.

However, the Church does state that censorship can be utilized in very extreme scenarios. Naturally, people will use this tool for harm. One of the major negative uses of social

---

http://www.vatican.va/roman_curia/pontifical_councils/pccs/do cuments/rc_pc_pccs_doc_22021997_ethics-in-ad_en.html.

[55] Pontifical Council for Social Communications, *Ethics in Internet* (February 22, 2002), 11, accessed August 23, 2019, http://www.vatican.va/roman_curia/pontifical_councils/pccs/do cuments/rc_pc_pccs_doc_20020228_ethics-internet_en.html #SOME%20AREAS%20OF%20CONCERN.

media in the modern world is the exploitation of children online. According to the National Center for Missing and Exploited Children, since 2002, which is when this pontifical document was issued, there have been more than 293 million images and videos of child abusive images online and more than 17,500 child victims.[56] These images can be used by child predators and sex traffickers. Again, the Church does see a need to protect the dignity of each human person, so if these children are being exploited, it is the responsibility of the government to censor such images.

Current State of Social Media Use

There are various websites that are being used by schools to help enhance lessons. In Catholic schools, formed.org, is being utilized for children. This website has a section for youth, which includes videos that correspond to the faith. There is an animated Bible series which portrays well-known Bible stories (i.e. Adam and Eve, Noah's Ark, the Good Samaritan) with modern graphics. There is also a cartoon series called "Brother Francis" in which this Franciscan brother talks about important aspects of the Catholic faith, including the seven Sacraments and the holy days of Christmas and Easter. Additionally, this website offers a study section for youth that relates more to a religious education program. Topics are about one's belief in God and prayer.

One of the biggest changes in education in the past decade has been the emphasis of character education. Books such as Sean Covey's *The 7 Habits of Happy Kids* have inspired teachers to start character education lessons, which instill in

---

[56] "Key Facts," *National Center for Missing and Exploited Children,* accessed August 23, 2019, http://www.missingkids.com/footer/media/keyfacts.

children positive traits to abide by, such as honesty, generosity and integrity. A website that is used is passiton.com from The Foundation for a Better Life. This website creates various commercials that range from thirty seconds to ninety seconds that "promote positive role models and remind people of the values we all share."[57] Although the website is not affiliated with any religious organization, it can be utilized in a Catholic elementary school. For example, if there is a lesson about the fruits of the Spirit, "love, joy, peace, forbearance, kindness, goodness, faithfulness, gentleness and self-control" (Gal 5:22), there are videos on this website about love, kindness and peace. Similarly, if there is a lesson about the sacrament of reconciliation, there is a video about the importance of forgiveness. Typically, in a standard lesson plan format, these videos from this website could be used during the motivational component, which is at the beginning of the lesson and is meant to excite the students and pique their interest.

One of the most popular websites being used by Catholic teachers is catholicmom.com. While this site contains a lot of valuable resources for Catholic women, there is a section that is dedicated to children and education. There are over fifty free lesson plans that are available to teachers that relate to various aspects of the faith. The website does state, "Some of our lesson plans contain links to Christian, non-Catholic resources. When preparing your lessons, we urge you to inform yourself on the teachings of the Catholic Church."[58] This website is valuable for new teachers who may seem overwhelmed with designing lessons. The links and information are all provided, and teachers can even upload

---

[57] "About Us," *The Foundation for a Better Life*, accessed August 23, 2019, https://www.passiton.com/who-we-are.

[58] "Lesson Plans," *Catholic Mom*, accessed August 23, 2019, http://catholicmom.com/education-2/lesson-plans/.

some of their own lessons. The disclaimer at the top is important because some Christian denominations have different teachings about certain aspects of the faith, such as the sacraments, so this notice is there to inform people that one should always return to the doctrine of the Catholic faith when implementing their lesson plans.

Pinterest is a teacher's dream website. It acts like a social forum where people, once they create an account, can post different ideas to this website with links to the actual sources. School teachers can type into the search bar any topic they need a lesson on, such as a specific saint or holiday. From there, various pictures and resources will appear with descriptions of the desired search. Once somebody clicks on the picture, it will take them to the actual website where the resource can be found. Again, this website is a valuable tool for new teachers. There are hundreds of activities, crafts, and lesson plans that can be used in the classroom. Sometimes, the person must subscribe to a website to gain access to the resources, but Pinterest is a great way to brainstorm new and engaging lessons. In 2012, *Forbes Magazine* commented on Pinterest saying, "Pinterest is a great looking site. It's an inviting platform with an exceptional user experience."[59]

Conclusion

Education in the United States is currently in a transition stage. In the state of New York, there are high-stakes tests which are changing the way curriculum is being taught. In order to compete with foreign countries, school admini-

---

[59] Lisa Arthur, "Pinterest: The Good, the Bad, and the Ugly," *Forbes,* April 3, 2012, accessed August 23, 2019. https://www.forbes.com/sites/lisaarthur/2012/04/03/pinterest-the-good-the-bad-and-the-ugly/#7b4a555f326f.

strators are putting pressure on their teachers to teach a certain amount of material, specifically mathematics and English language arts, and finish that curriculum within the academic year, while the students must take state tests that reflect the material that was taught. There are several issues that arise from this startling scenario. First, students are only being taught material that will be on the test. This puts pressure on teachers to complete a certain amount of work by the state test dates; otherwise, their students are placed at an unfair advantage. Second, because state governments, and to some extent the federal government, dictate what curriculum must be taught, teacher creativity and ingenuity are stifled. Every child in each state, such as New York, is being taught the same curriculum. While this may sound wonderful that every child is learning the same lessons, this does not allow different cultural and ethnic perspectives and customs to be brought into the lessons. Hence, a child in Cooperstown will have the same reading literature as a child in South Ozone Park, without any consideration as to the cultural backgrounds of a given school. That is ironic since the culture is pushing for cultural diversity, not cultural plurality. Third, other school subjects are not taught with as much rigor as the ones mentioned above. Therefore, in Catholic schools, while religion should be the focus, it is placed on the "backburner" because there are more important tasks that must be fulfilled.

While the United States is still in transition, debating which mode of teaching is more effective—the extremes being an overly academic program to a Montessori method of teaching—Catholic schools would have an opportunity to use a new Catholic educational website that would be fun for the students and a great resource for the teachers. As stated at the beginning, one outcome would be to have half of all Catholic elementary schools within a given diocese incorporate this website by the end of the academic year. One way to promote this website would be by contacting the diocese to pitch the idea and ask if it is permissible to post a link to the website on the diocesan website and Facebook page. This would be done in June before the new academic year. From that initial post, one repost would be done every week throughout the summer, until two weeks before school begins. At that time, a post of the new website will be made every three days, to prompt teachers to use it. It is typically within the last three weeks of the summer that teachers begin to prep work for the classroom. Finally, with the permission of the diocese, one week before the initial school year begins, I would send out an e-mail blast to all the Catholic elementary schools, with a link to the website. Included on that e-mail would be a promotional offer for schools to subscribe to the website by September 1 of that school year. The number of teacher subscribers would also be important. Hopefully, by the end of the academic year, June 30, over 150 teachers would be registered with the website. Including a promotional offer of a discount for first year subscribers will be a great way to entice teachers to sign-up.

Over the course of three years, I would hope to see the use of the website grow from fifty percent of Catholic elementary schools in a diocese to eighty percent of Catholic schools in a diocese. If that is the case, then in the third year, I hope to pilot the website in two neighboring dioceses. When I approach these dioceses, I hope to have substantial statistics, such as success rates of student performances and reviews from teachers and schools using this website. A new video can be created which showcases teachers and students using the website and highlighting which parts of the website are most engaging.

4

Encouraging Catechism Students to Become More Involved in Parish Activities

Jinwoo Nam, M.Id., Seminarian, South Korea, Idente Missionaries

The sixth graders in the religious education program of our Lady of Solace-St. Dominic Parish in the Bronx, New York, are the students whom I am going to catechize in the year 2019-20. Based on my observations during my four years of experience as the sixth-grade catechist, almost of all the students go to public school. Each year there is an average number of twenty students in the catechism class. Most of them will receive the Sacrament of Confirmation in the following year. In my experience, after they receive Confirmation, most stop participating in the Holy Mass and parish activities. Therefore, I find it important that students in the religious education program begin to engage in the youth group or various services so that they will joyfully persevere in their faith in their future. In previous years, I experienced their expressions of indifference, disconnection, and boredom about Catholicism in my class. I hope that the blog, my social media project, will interest my students in becoming involved in the youth group, called *Idente Youth,* or various services in the parish by the end of the year.

YOUNG DISCPLES 6
Save the Church. Salvad la Iglesia.

Welcome to our Blog

The blog that I created in this project is online at https://www.sundays.school.blog/. I am going to use this blog to communicate and share with my catechism students various friendly content related to our faith. I posted the welcoming message to introduce myself and the blog and to share its purpose with my students. By uploading content such as videos, pictures, news articles, gospel passages, and inspirational texts, I am going to engage my students in our faith. The students can share their views and comments with me and each other.

The Catholic Church is aware of the good and the evil associated with the use of social media. She affirms social media as a gift of God that provides guidance and unites mankind. Pope Pius XII states in his encyclical *Miranda Prorsus,* "those very remarkable technical inventions which are the boast of the men of our generation, though they spring from human intelligence and industry, are nevertheless the gifts of God."[60] The ultimate origin of social media is truly God Himself, though it is not without human achievement. God the Father bestows us with social media, taking continuous care of His children.[61] He wants to bring us into deeper communion and help us to be saved in accordance with His will.[62]

Deeper communion or mutual understanding is possible with social media. According to Saint Pope John Paul II's speech for the 28th World Communications Day, "Television can enrich family life. It can draw family members closer together and foster their solidarity with other families and with the community at large."[63] Television is an example of how media can foster common knowledge and enrich the culture of solidarity. Therefore, media helps us to understand each other as sisters and brothers beyond the national borders. John Paul II affirms that a broad range of communication through media is quite helpful for evangeli-

---

[60] Pius XII, Encyclical Letter of His Holiness On Motion Pictures, Radio & Television *Miranda Prorsus* (September 8, 1957), accessed August 18, 2019, http://w2.vatican.va/content/pius-xii/en/encyclicals/documents/ hf_p-xii_enc_08091957_miranda-prorsus.html, 1.

[61] Pius XII, *Miranda Prorsus, 2.*

[62] Pontifical Council for Social Communications, Pastoral Instruction on the Means of Social Communication Written by Order of the Second Vatican Council *Communio Et Progressio* (May 23, 1971), accessed August 18, 2019, http://www.vatican.va/roman_curia/pontifical_councils/pccs/documents/rc_pc_pccs_doc_23051971_communio_en.html, 2.

[63] John Paul II, Message of the Holy Father John Paul II for the 28th World Communications Day (January 24, 1994), accessed August 18, 2019, https://w2.vatican.va/content/john-paul-ii/en/messages/communications/documents/hf_jp-ii_mes_24011994_world-communications-day.html, 2.

zation, noting that, "It can increase ... their religious knowledge, making it possible for them to hear God's word, to strengthen their religious identity, and to nurture their moral and spiritual life."[64] For the Catholic Church, it can be a tool that unites Catholics across the entire world and helps us to share the teaching of Jesus with others who do not know Him.

Social media is conducive to instruction as well. Saint Pope Paul VI acknowledges, "the media of social communication have this educational capacity that the Church proposes."[65] According to John Paul II's letter *The Rapid Development*, "the communications media have acquired such importance as to be the principal means of guidance and inspiration for many people in their personal, familial, and social behavior."[66] In a world globalized with the help of the technological advancement of media, the media provides not only entertainment but also guidance and inspiration to help people become more virtuous. Therefore, social media can be a great instrument to instill positive values in individuals, family and society.

The Church is aware of the evil to avoid as well as the good to pursue. The Church recognizes that media can be used against God's plan and humanity.[67] Paul VI notes that a great

---

[64] John Paul II, Message of the Holy Father John Paul II for the 28th World Communications Day, 2.

[65] Paul VI, Message of the Holy Father for the World Social Communications Day (April 11, 1976), accessed August 18, 2019, ahttps://w2.vatican.va/content/paul-vi/en/messages/communications/documents/hf_p-vi_mes_19760411_x-com-day.html, 4.

[66]John Paul II, Apostolic Letter *The Rapid Development*, accessed August 18, 2019, https://w2.vatican.va/content/john-paul-ii/en/apost_letters/2005/documents/hf_jp-ii_apl_20050124_il-rapido-sviluppo.html, 3.

[67] Vatican Council II, Degree on the Media of Social

deal of harm occurs when men produce media in defiance of honest, moral responsibility.[68] Indeed, social media have been an express instrument of propagating morally degrading messages. Saint John Paul II also states that television manipulates people "by propagating degrading values and models of behavior" and "inculcating moral relativism and religious skepticism."[69] For example, according to *Ethics in Internet*, internet facilitates the transmission of western secularism, such as the disunity of marriage and family, to the rest of the world because of the cultural domination of the west.[70] Additionally, *Ethics in Advertising* says, "it [advertising] also can be vulgar and morally degrading. Frequently it deliberately appeals to such motives as envy, status seeking and lust."[71] Some communicators of social media use violence and sex to gain their profits regardless of how these affect

---

Communications *Inter Mirifica* (December 4, 1963), accessed August 18, 2019, http://www.vatican.va/archive/hist_councils/ ii_vatican_council/documents/vat-ii_decree_19631204_inter-mirifica_en.html, 2.

[68] Paul VI, Message of the Holy Father for the World Social Communications Day (May 7, 1967), accessed August 18, 2019, http://w2.vatican.va/content/paul-vi/en/messages/ communications/documents/hf_p-vi_mes_19670507_i-com-day.html, 6.

[69] John Paul II, Message of the Holy Father John Paul II for the 28th World Communications Day, 3.

[70] Pontifical Council for Social Communications, *Ethics in Internet* (February 22, 2002), accessed August 18, 2019, http://www.vatican.va/roman_curia/pontifical_councils/pccs/do cuments/rc_pc_pccs_doc_20020228_ethics-internet_en.html, 11.

[71] Pontifical Council for Social Communications, *Ethics in Advertising* (February 22, 1997), accessed August 18, 2019, http://www.vatican.va/roman_curia/pontifical_councils/pccs/do cuments/rc_pc_pccs_doc_22021997_ethics-in-ad_en.html, 13.

people in society.

Although social media can be used to promote brotherhood and mutual understanding, it may give rise to division among men as well. According to *Ethics in Advertising*, some communicators are reluctant to address the educational and social needs of particular groups in society, such as children, the elderly and the poor, for they want to remain comfortable within themselves.[72] John Paul II adds that advertising in the media creates prejudice against certain groups that can, for example, lead to the mistreatment of women, which is contrary to human dignity. Pope Francis also said on the 53rd World Communication Day, "in the social web identity is too often based on opposition to the other, the person outside the group."[73] Social media can be used as a space of disputation, hatred and envy, instead of promoting brotherhood.

Furthermore, social media may give out inaccurate information. John Paul II observes television may "spread distorted, manipulative accounts of news events and current issues."[74] Some producers of social media change information in order to attract more customers and gain more profits. *Ethics in Advertising* mentions consumerism as a trigger of the occurrence of inaccurate information in media, saying the "competitiveness and round-the-clock nature of Internet journalism also contribute to sensationalism and rumor-mongering, to a merging of news, advertising, and

---

[72] Pontifical Council for Social Communications, *Ethics in advertising*, 12.

[73] Francis, Message of the Holy Father Francis for the 53rd World Day of Social Communications Francis, accessed August 21, 2019, https://press.vatican.va/content/salastampa/en/bollettino/pubblico/2019/01/24/190124b.html.

[74] John Paul II, Message of the Holy Father John Paul II for the 28th World Communications Day, 3.

entertainment, and to an apparent decline in serious reporting."[75] Because there is so much information on the internet, it is difficult to check the truthfulness and appropriateness of information.[76] Our moral responsibility to share the truth is often ignored because of greed.

The Church encourages her members to use social media by taking advantage of the good and avoiding the evil that it causes. As mentioned above, social media is a gift of God for the promotion of brotherhood and the guidance of His children. The Church wants people to use social media to unite people and guide them to salvation, which is their destiny. Therefore, according to *The Church and Internet*, priests, deacons, religious and lay pastoral workers have to go through training about media.[77] Moreover, the Second Vatican Council states in *Inter Mirifica* that it is the responsibility of pastors to assist people in the use of social media for evangelization.[78] *Aetatis Novae* further states, "Every episcopal conference and diocese should ... develop an integrated pastoral plan for communications, preferably in consultation with representatives of international and national Catholic communications organizations and with local media professionals."[79] Likewise, the Church promotes the training and

---

[75] Pontifical Council for Social Communications, *Ethics in advertising*, 13.

[76] Ibid.

[77] Pontifical Council for Social Communications, *The Church and Internet* (February 22, 2002), accessed August 18, 2019, http://www.vatican.va/roman_curia/pontifical_councils/pccs/do cuments/rc_pc_pccs_doc_20020228_Church-internet_en.html, 7.

[78] Vatican Council II, *Inter Mirifica*, 3.

[79] Pontifical Council for Social Communications, Pastoral Instruction on Social Communications on the Twentieth Anniversary of Communio et Progressio *Aetatis Novae* (February

use of social media for the purpose of evangelization at the local, diocesan and national levels. Every member of the Church should know how to responsibly and smartly achieve the mission of the Church by means of the internet.[80]

At the same time, the Church warns us to be responsible to avoid the evil that social media causes. As mentioned earlier, social media sometimes brings about division and prejudice among people, propagation of moral relativism, and inaccuracy of information. The Church does not shy away from technology because of the negative effects of the internet but rather is willing to face challenges and learn to make positive use of the internet.[81] In order to deal with division, the Church always promotes unity and reconciliation because forms of media are "agents of unity, binding all Christians together in solid and visible charity, in the harmony of a single purpose, in a firm orientation towards goodness and towards God."[82] Social media must be a positive means toward God in communion with one another. Paul VI said on the first World Communication Day, "The greater, in fact, the power and the ambivalent efficacy of these means, the greater must be the care and the responsibility with which they are to be

---

22, 1992), accessed August 18, 2019, http://www.vatican.va/roman_curia/pontifical_councils/pccs/documents/rc_pc_pccs_doc_22021992_aetatis_en.html, 23.

[80] Pontifical Council for Social Communications, *The Church and Internet*, 10.

[81] Ibid., 10.

[82] Paul VI, Message of the Holy Father for the World Social Communications Day (April 19, 1975), accessed August 18, 2019, https://w2.vatican.va/content/paul-vi/en/messages/communications/documents/hf_p-vi_mes_19750419_ix-com-day.html, 2.

employed."[83] Powerful and developed countries must be responsible to employ social media for the good of developing countries. In opposition to prejudice, the Second Vatican Council states that the Church encourages communicators to fairly provide the content of social media based on the needs of groups, especially the youth, in terms of cultural background and ages.[84] Concerning inaccuracy of information, the Church advises producers of media to take seriously the responsibility to communicate the truth for the growth of men's freedom.[85] Therefore, when they provide public opinions, they must be honest and truthful. As John Paul II says, "the fundamental moral requirement of all communication is respect for and service of the truth."[86] Communication of the truth is for the common good and for the Kingdom of God.

Social media, a new form of media today, is different from the traditional media, such as television, radio, and the press. McAfee mentions, "This media is limited where producers will

---

[83] Paul VI, Message of the Holy Father for the World Social Communications Day (May 7, 1967), accessed August 18, 2019, http://w2.vatican.va/content/paul-vi/en/messages/communications/documents/hf_p-vi_mes_19670507_i-com-day.html, 7.

[84] Vatican Council II, *Inter Mirifica*, 16.

[85] John Paul II, Message of the Holy Father John Paul II for the 20th World Communications Day (May 11, 1986), accessed August 18, 2019, at https://w2.vatican.va/content/john-paul-ii/en/messages/communications/documents/hf_jp-ii_mes_24011986_world-communications-day.html, 18.

[86] John Paul II, Message of the Holy Father John Paul II for the 37th World Communications Day (June 1, 2003), accessed August 18, 2019, at https://w2.vatican.va/content/john-paul-ii/en/messages/communications/documents/hf_jp-ii_mes_20030124_world-communications-day.html, 3.

create, package, and send information to users who are fed that information which was seen as fit to consume."[87] In the traditional media, a producer creates media, and consumers receive it. On the other hand, the new media, called *social media*, such as Facebook, Instagram, Twitter, YouTube, and Snapchat, does not limit the production of consumers. [88] It permits the users to produce and exchange their ideas quickly, even across borders, and at low costs.[89] For example, it is possible to create one's own YouTube channels to take and share videos through smartphones in the place of TV channels. Or, it is possible to share one's own news in Instagram through smartphones, instead of relying on the newspaper. Thus, elimination of walls between producers and consumers makes the new media "social."[90]

The youth are always the generation that is most sensitive to new development of social media.[91] They depend on smartphones and social media. Jean M. Twenge coined the term *iGen* to describe this generation, which is fully immersed in the use of smartphones in the period of adolescence. They were born between the years 1995 and 2012. The audience of my project (6th graders in 2019) belongs to *iGen*. Twenge mentions, for *iGen*, "time spent with friends in person has been replaced by time spent with friends (and virtual friends) online."[92] They are more comfortable with using smartphones

---

[87] S.A. McAfee, *Social Media Magisterium* (St. Louis, MO: En Route Books and Media, 2018), 26.

[88] Ibid.

[89] Ibid.

[90] Ibid., 27.

[91] Ibid., 17.

[92] Jean M. Twenge, *iGen: Why Today's Super-Connected Kids Are Growing Up Less Rebellious, More Tolerant, Less Happy--and Completely Unprepared for Adulthood--and What That Means for the Rest of Us*, Second Edition (New York: Atria Books, 2017), 73.

and internet. This also means that they are experts in the use of smartphones and internet. At the same time, Twenge's study shows that activities such as gaming, social media, TV, etc. makes a person unhappy, depressed, and lonely while activities such as sports, homework, in-person encounter, etc., make a person happier.[93]

On the basis of the social trend of the increasing use of social media, different religions are trying to utilize social media to spread their religions. Many of them use Facebook to share their religious beliefs, for example, by posting inspirational quotes from popular Church leaders, employing different approaches to missionary work, such as posting memes containing religious themes and creating prayer networks.[94] In the interviews of Lugard Nduka and John McGuire in 2017, the faithful benefit from engaging in the religious life on social media, such as Facebook and Twitter during the weekdays instead of only the weekend.[95] Nduka and McGuire recommend educating the faithful about faith through online, "Sunday Mass homilies could be videotaped and uploaded onto the parish or diocese website. It would give Catholics of all ages a second (or perhaps first) chance of

---

[93] Twenge, *iGen*, 77-8.

[94] Kris Boyle, Jared Hansen, and Spencer Christensen, "Posting, Sharing, and Religious Testifying: New Rituals in the Online Religious Environment," in *Religion Online: How Digital Technology is Changing the Way we Worship and Pray*, eds. August E. Grant, Amanda F.C. Sturgill, Chiung Hwang Chen, and Daniel A. Stout (Santa Barbara, CA: Praeger Publishers, 2019), 15.

[95] Lugard Nduka, and John McGuire, "The Effective Use of New Media in Disseminating Evangelical Messages among Catholic College Students," *Journal of Media and Religion* 16, no. 3 (2017): 101, accessed October 4, 2019, ATLA Religion Database with ATLASerials PLUS.

hearing instruction."[96] From what Inaku K. Egere says, priests can use Facebook or Twitter to share "short daily reflections, pictures of holy symbols, daily readings, catechesis and even online retreats, upcoming events, barns, Easter/Christmas messages, notices, etc."[97] He says that these will help people to be close to one another and to be closer to God in a world that is getting busier.[98] He insists that the Church should learn to adopt the marketing strategies of businessmen in the secular world who know how to attract people to buy their projects.[99] Likewise, the Church should attract people to receive the Good News.

This current state of media, described above, encourages me to make proper use of social media for evangelization. I intend to create and operate a blog as an online space of the community of my catechism students. I want to reach out to *iGen*, which feels comfortable with smartphones, by making the Good News easily accessible and attractive to them. In fact, two blogs have inspired me. They are Life Teens Blog (https://lifeteen.com/blog/), and the Blog of St. Joseph Worker Parish in Iowa (https://theworker.org/blog/). Life Teens Blog is sensitive to and friendly with the youth in terms of its themes of articles and vocabularies. For example, one article, written by Emily Clare Burt, "Good Morning Beautiful: Glorify in Your Morning Routine," provides practical guidelines about physical, mental and spiritual health. The Blog of St. Joseph Worker Parish in Iowa creates an intimate

---

[96] Ibid.

[97] Inaku K. Egere, "Social Media and Mission-Based Marketing Approach for New Evangelization in the Digital Age African," in *African Ecclesial Review* 57, nos.3&4 (2015): 194, accessed August 21, 2019, ATLA Religion Database with ATLASerials PLUS.

[98] Egere, "Social Media and Mission-Based Marketing Approach," 194.

[99] Ibid.

space for the parish community, offering the news of the parish, short inspirational sentences, parish projects, hymns of the Mass and job postings. The Blog also plays a role of encouraging parishioners to get together, for example, by announcing a group discussion of a Catholic book in the parish. The Blog of St. Joseph Worker Parish is like my project in terms of creating a communal space for a specific audience.

In conclusion, I expect to measure the success of my articulated outcome by keeping track of the number of new members from my catechism class who become involved in parish activities, such as the youth group called *Idente Youth*, altar servers, lectors, ushers, and choir members. One's involvement in the parish activities is a sign of his or her commitment and love for Jesus Christ and for the Church. In the recent catechetical years 2017-18 and 2018-19, there was only one new committed member of parish activities from my catechism class (which includes, on average, twenty students per year). My goal for this year, with the help of the new blog, is to involve three new committed members in one of the parish activities. In order to build on that success over time, I will consistently post new content on a weekly basis and prepare the content at least one month earlier. Following the examples of Egere, I am going to engage my students during weekdays with short reflections, pictures, catechesis upcoming events and inspirational messages. I am going to advertise the parish youth group and other parish activities by posting pictures or videos in the blog and in the class. The blog will connect them to the Facebook and Instagram pages of the youth group. I am going to encourage my students to confidently share their comments and exchange their views about the content.

Social media is neither entirely good nor evil. It depends on how we use it. The Church encourages us to make positive use of social media for evangelization. The Church recognizes it as the gift of God. It creates opportunities to unite all of us together and provide inspiration and guidance in our lives. However, we must avoid using it to manipulate people, falsify the truth, degrade morality and divide people. If we use social media for Christ, Christ will use our efforts to unite and sanctify mankind. Nowadays, social media is a kind of language and space that especially the youth, iGen, use. Just as we learn a new language to talk with people living in another country, we can learn the language of iGen. This concept is not new at all. Saints and missionaries of the Church did this to evangelize in foreign lands and cultures while holding onto the morality and faith of the Church. If we insist on only traditional methods of evangelization, we may miss people God the Father expects us to take care of. Many organizations and individuals have realized the need to evangelize through social media. Other religions use it as well. People find religious content in social media helpful for living their spiritual lives. The time to wait for people to come to the Church is coming to an end. It is time to reach out to the places where they are. Where are our catechism students during the week? They are in social media and with smartphones. They can encounter Jesus through social media and smartphones.

5

What Being in Love Means

Tobechukwu J. Offiah, Seminarian, Nigeria,
Diocese of Brooklyn

Introduction

To love and to be loved remains an existential part of human life, especially as it is practiced in the relationship between a man and a woman. Social media today contain much information about love and relationships. Some of this information gives a false view of what it means to love. Some see love as sexual relations, whereas others consider love primarily in terms of how it benefits them. Considering this, this work of mine joins its voice with those that seek to educate people on what "true love" really means. It has as its audience Millennials and Generation Z, especially those on Facebook. The choice of audience is based upon the fact that this is the age bracket when people desire to enter into a relationship, and knowing very well that if they get it right, they can model for their children what it means to be in a successful relationship.

The Artifact

The artifact below portrays the true meaning of love. It derives its meaning from the perfect love of Christ "who, being in the form of God did not count equality with God something to be grasped. But he emptied himself, taking the form of a slave, becoming as human beings are" (Phil 2: 6-7; New Jerusalem Bible). A mere look at the artifact shows a man and a woman holding hands, but an in-depth study reveals an act of emptying by both parties and an act of filling by the other. True love lies in the self-emptying and the giving of oneself for the *good* of the other. For without the emptying of either hand, it becomes difficult for both hands to come together in a full expression of love.

True Love Is the emptying of one's self, in order to be filled by the other.

Church Teaching of Evangelization in Social Media

One cannot speak about the Church without talking about communication. This is because communication has been central to the mission of the Church since its inception. According to the Pontifical Council for Social Communications, "Communication in and by the Church finds its starting point in the communion of love among the divine Persons and their communication with us."[100] This to a great extent shows how important communication is in the life of the Church. For this reason, the magisterium has written many documents on the means of communication (especially social media) and how we may use it in line with the divine plan, which is aimed toward the glorification of God and the salvation of souls. For the purposes of this work, these documents shall be broken into three main sections, namely, the power of the media, the role of the members of the Church and the methodology of using the media for evangelization.

The Power of the Media

There is no doubt that the advent of social media has brought about tremendous development in our lives. Information can now be easily sent out to the rest of the world just by the click of a button; making business transactions and acquiring professional knowledge from desirable educational institutions across the globe has become very easy, thanks to the invention of social media. The Church acknowledges these and many other benefits. Thus, at the beginning of *Inter Mirifica,* the Second Vatican Council calls attention to the importance of social media, noting that "The Church recognizes that these media, if properly utilized, can be of great service to mankind, since they greatly contribute to men's

---

[100] Pontifical Council for Social Communications, *Ethics in Communications.* (June 4, 2000). n. 3. http://www.vatican.va/roman_curia/pontifical_councils/pccs/documents/rc_pc_pccs_doc_20000530_ethics-communications_en.html

entertainment and instruction as well as to the spread and support of the Kingdom of God."[101] By this, the Church observes the significant roles social media play in the lives of people. They can be utilized to build up excitement as well as to inform or become informed about current events. The Church also recognizes that social media can play an essential role in spreading the good news of salvation and supporting the work of God. This is the reason Pope Pius XII in his encyclical letter *Miranda Prorsus* (1957) postulates, "Motion Pictures [social media], which came into existence some sixty years ago, must today be numbered among the most important means by which the ideas and discoveries of our times can be made known."[102]

Nevertheless, the Church also calls our attention to the fact that the media, which has many benefits, can also be put to negative uses. Thus, she asserts, "the Church recognizes, too, that men can employ these media contrary to the plan of the Creator and to their own loss."[103] In other words, the Church postulates that social media, which on the one hand can help us draw closer to God, can also be used to our detriment if it distances us from God. This is the reason Pope Paul VI cries out in his message for the 4th World Social Communications Day that "we witness young people and children... being dragged into the pit-falls of eroticism and violence or led along

---

[101] Vatican Council II, Decree on the means of social communication *Inter Mirifica*. (December 4, 1963), 2. https://www.vatican.va/archive/hist_councils/ii_vatican_council /documents/vat-ii_decree_19631204_inter-mirifica_en.html

[102] Pope Pius XII, Encyclical Letter *Miranda Prorsus*. (September 8, 1957), accessed on September 03, 2019, http://www.vatican.va/content/pius-xii/en/encyclicals/ documents/hf_p-xii_enc_08091957_miranda-prorsus.html.

[103] Vatican Council II, *Loc. Cit.*

the perilous paths of incertitude, anxiety and anguish"[104] as a result of the misuse of the social media.

The Role of the Members of the Church

Given these adverse uses of the media and the negative effects they have, is it still advisable for members of the Church to be involved in it? To this end, Pope John Paul II posits, "do not be afraid of new technologies (social media)!... which God has placed at our disposal to discover, to use and to make known the truth, also the truth about our dignity and about our destiny as his children, heirs of his eternal Kingdom."[105] In like manner, the Church, through the Pontifical Council for Social Communications, now operating under the name Dicastery for Communications, encourages its members that "hanging back timidly from fear of technology (social media) or for some other reason is not acceptable, in view of the very many positive possibilities of the Internet."[106] Thus, instead of completely rejecting the use of social media because of its possible dangers, the Church sees the enormous good that can be achieved through it,

---

[104] Paul VI, "Social Communications and Youth." (April 6, 1970). http://www.vatican.va/content/paul-vi/en/messages/ communications/documents/hf_p-vi_mes_19700406_iv-com-day.html

[105] John Paul II, *The Rapid Development*, Apostolic letter to those responsible for communications. (January 24, 2005). n. 14. http://www.vatican.va/content/john-paul-ii/en/apost_letters/ 2005/documents/hf_jp-ii_apl_20050124_il-rapido-sviluppo.html

[106] Pontifical Council for Social Communications. *The Church and Internet*. (February 22, 2002). n. 10. http://www.vatican.va/ roman_curia/pontifical_councils/pccs/documents/rc_pc_pccs_d oc_20020228_church-internet_en.html

which also includes helping to correct the misuse.

Furthermore, *Aetatis Novae* states that "Against this background we (the Church) encourage the pastors and people of the Church to deepen their understanding of issues relating to communications and media, and to translate their understanding into practical policies and workable programs,"[107] for "a merely censorious attitude on the part of the Church toward the media is neither sufficient nor appropriate."[108] By this the Church calls on all her members from the clergy down to the laity to first deepen their knowledge of communication and social media. This is because no one gives what he does not have *(Nemo dat quod non habet)*. For the members of the Church to be able to evangelize through the media, they must have enough knowledge about the media. Nevertheless, from the above, one will observe that the Church does not just want her members to acquire knowledge about the media, but also to be practically involved in the affairs of the media, to be a light unto souls and to dispel the darkness of misuse inherent in the media.

It is to this end that the Church calls on her pastors "to have a sound knowledge of all questions which confront the souls of Christians with regard to Motion Pictures, Radio and

[107] Pontifical Council for Social Communications. *Aetatis Novae*. Pastoral Instruction on Social Communications on the Twentieth Anniversary of "Communio Et Progressio". (February 22, 1992). n. 3. http://www.vatican.va/roman_curia/pontifical_councils/pccs/documents/rc_pc_pccs_doc_22021992_aetatis_en.html

[108] Pontifical Council for Social Communications, *Pornography and violence in the communications media: A pastoral response.* (May 7, 1989), n. 30. https://www.vatican.va/roman_curia/pontifical_councils/pccs/documents/rc_pc_pccs_doc_07051989_pornography_en.html

Television (social media);"[109] additionally, they should "use these aids correctly as often as, in the prudent judgment of ecclesiastical authority, the nature of the ministry entrusted to him [them] and the need of assisting an increasing number of souls demand it."[110] In addition to acquiring knowledge and using the media in accordance to ecclesiastical authority, it is also "the duty of Pastors to instruct and guide the faithful so that they, with the help of these same media, may further the salvation and perfection of themselves and of the entire human family."[111] This reveals that the function of evangelization through the media is not the sole responsibility of the priest; rather, it is a call to action to every member of the Church, both clergy and lay. Nevertheless, the priest has a duty to instruct the faithful on the media and how to fruitfully utilize it for the work of salvation. As for the laity, they are called to "strive to instill a human and Christian spirit into these media, so that they may fully measure up to the great expectations of mankind and to God's design."[112] Thus, the laity on their part also have important roles to play in this evangelizing mission of the Church through the media.

Methodology of Evangelization through the Media

The Church in her wisdom does not just ask her members to evangelize through the media; she also offers some guidelines. Nonetheless, before they evangelize through social media, Church members should be equipped with the laws of morality. Thus, the Church asserts, "in both the search for news and in reporting it, there must be full respect for the laws

---

[109] Pope Pius XI, *Loc. Cit.*
[110] Ibid.
[111] Vatican Council II, *op. cit.* n. 3.
[112] Ibid.

of morality and for the legitimate rights and dignity of the individual."[113] Having the law of morality, the rights and dignity of the human person and the fear of God as her backbone, the Church through the Pontifical Council for Social Communications has encouraged her members to use two basic methodologies in participating in the media: the research phase and the design phase.

The research phase involves identifying the areas of ministry requiring attention, be it in the parish or the society. Secondly, it involves considering what is currently being done (including its effectiveness) in the area of ministry identified and knowing its strengths and weaknesses. Thirdly, it entails identifying communications resources, technology, and personnel available to the individual.[114]

The design phase involves using the information gathered and the resources, technology and personnel available to the individual to publicize information in the media. *Aetatis Novae* identifies various fields, such as education, spiritual formation and pastoral care, cooperation, communications and development of peoples, etc., in which this may be achieved.[115]

Current State of Social Media

Looking at our society today, one hears the echo of love resounding everywhere. It can be heard in the family, in the Church, in relationships, etc. Nevertheless, what this love means is comprehended differently by different individuals or groups. This to some extent has informed what people put out

---

[113] Ibid, n. 5.

[114] Pontifical Council for Social Communications. *Aetatis Novae. Op. cit.* n. 26.

[115] Ibid. n. 27-33

in social media today. The media is full of references to and images of the love that should exist in a male and female relationship. Nonetheless, not all these works portray the true meaning of love. Some see love from the point of view of sensual relationships; some from the viewpoint of the benefits to be received from the other; some from the perspective of possessiveness. This is one of the reasons Prince, a YouTuber, asserts that "who you hate is probably somebody you use to love...how can love ever turn to hate in a single instance, was that love?"[116]

The media have made many attempts to correct this false view of love. The website of the United States Conference of Catholic Bishops offers many teachings regarding love and relationships. For example, the United States Conference of Catholic Bishops affirms that "God desires that each of us grow in holiness by giving and receiving love like him. Such love requires putting the human ego aside and first considering the needs of others before our own."[117]

"St. John Paul II Catholic Newman Center," which has as its mission to help souls "know the depth of God's love and a school to form authentic witnesses to love truth, and life,"[118] has also done a lot of work in this regard. They have written articles, interviews, made videos, etc. of what it means to love. On their website, there is a quote by Pope John Paul II that

---

[116] Prince Ea, *Love (Everything you have been told was a lie)*, YouTube video posted on June 17, 2014. accessed on September 7, 2019, at www.youtube.com.

[117] United States Conference of Catholic Bishops, *Love and Sexuality*. Accessed on September 7, 2019. http://www.usccb.org/beliefs-and-teachings/what-we-believe/love-and-sexuality/index.cfm#communion

[118] St. John Paul II Catholic Newman center, *Our Mission*. Accessed on September 7, 2019. http://www.isucatholic.org/our-mission.html

"only the chaste man and the chaste woman is capable of true love."[119] This goes to portray that there is no necessary connection between love and sensual relation.

Mary Beth Bonacci wrote an article on EWTN entitled "Expressing love in a relationship." Here, she addressed questions relating to love in a relationship. According to her "love, real love, is above all wanting what's best for the other person. It's really caring about that person's well-being. Love means never, ever unnecessarily putting that person at any kind of risk;"[120] be it physically, emotionally, spiritually, etc.

From the above, it can be seen that some works have been done in educating people on what it means to love. Thus, this work is geared towards adding a voice to the already existing voices to spread the gospel of love far and wide.

Conclusion

Social media have proved to be of great importance to the Church and society at large. This is the reason the Church calls on "all the children of the Church ...(to) join, without delay and with the greatest effort in a common work to make effective use of the media of social communication in various apostolic endeavors, as circumstances and conditions demand."[121] In line with this call, this work finds a niche in educating its audience on what true love is through Facebook. For one to understand the meaning of love, one must first turn to Him who is love and who in time loved perfectly—Jesus. It

---

[119] St. John Paul II Catholic Newman Center, "Dating in College: Free to Love." Illinois State University, Diocese of Peoria. http://www.isucatholic.org/dating-and-relationships.html

[120] Mary Beth Bonacci, *Expressing love in a relationship,* at EWTN. Accessed on September 7, 2019. https://www.ewtn.com/catholicism/library/expressing-love-in-a-relationship-12662

[121] Vatican Council II, *op. cit.* n. 13.

is only by analyzing how he loved can we arrive at the true meaning of love. Thus, the author aims at providing educational content on the subject matter in the form of memes, articles, movies, music, etc. He also hopes to help the audience to enter a healthy relationship that will end up in Christian marriage. The audience includes Millennials and Generation Z who are currently using Facebook. The author intends to measure the success of the outcome of the work through feedback and comment left on the page.

Finally, this clarion call to participate in the media for evangelization is a pertinent one. If we shy away from the media because of some of its content, then the truth gradually fizzles away. But just as the Church fathers said "...communication of truth can have a redemptive power, which comes from the person of Christ;"[122] hence, we are called to spread the truth in every area of life in the media. Nevertheless, one should hearken unto the voice of the Church to be morally equipped as we participate in the media because "if the moral order is fully and faithfully observed, it leads man to full perfection and happiness."[123]

---

[122] Pontifical Council for Social Communications. *Aetatis Novae. Op. cit.* n. 6.

[123] Vatican Council II, *op. cit.* n. 6.

6

Social Media in Promotion Bible Study and Relationship

Anh Ngoc-Quoc Vu, Seminarian,
Diocese of Bridgeport, CT

Many years ago, when I first heard about social media like Twitter and Facebook, I thought that they were not for me. I could not imagine how people feel interested in using them. Nowadays, it is without a doubt that social media have strongly affected our lives in many different aspects. They use them to share pictures, music, spiritual reflections and political thoughts. I am working in St. Cecilia's Parish in Stamford, Connecticut, this year. One of the many activities that I get involved in is the Bible study class. We are studying the four Gospels and letters of St. Paul. There are about twenty members, and we gather once a week in the parish hall for the class. I realize that many of the students need to know beforehand the questions and suggestions related to topics in the class so that they have time to prepare before coming to the class. They also want to know the Sunday Gospel and want to share their experience and spiritual thought with others. For these reasons, I plan to use Facebook as a platform so that I can post questions of the class, the Sunday Gospel, some personal reflections and some suggestions from the Pope and bishops. Parishioners in my Bible study class can share them with other people in the parish and their friends. By doing this, I want parishioners in my parish to have more opportunities to evangelize the Word of God and to promote a good relationship with others.

The story of Martha and Mary is in the Gospel of Luke 10:38-42. Martha complained to Jesus that her sister Mary had left her to do all the serving. Jesus said to Martha that she worried too much about many things while Mary has chosen the better part by sitting and listening to Jesus. "Having Mary's heart in Martha's World" means that we still can spend time to listen to God's Word even though our lives are so busy with much work to do. Time spent with God is never wasted. Sharing this picture on social media to invite people to join the Bible study class is a way of evangelization and promotion of good relationships.

Having Mary's Heart in Martha's World

We now need to examine the Church teaching regarding evangelization in social media. Social media have become an important part of our daily lives. First, we need to know what "social media" mean. R. J. Dominick defines social media as "[o]nline communications that use special techniques that involve participation, conversation, sharing, collaboration, and linkage."[124] Meredith Gould also says, "Social media are web-based tools for interaction that, in addition to conversation, allow users to share content such as photos, videos, and links to resources."[125] More and more people continue spending time on social media each day for different reasons. Many people use social media to gain knowledge in study, to search for valuable information and daily news; other people use social media to share pictures and videos. Facebook is an example of the power of social media. In 2012, the number of active Facebook users was one billion. In the third quarter of 2019, it reached 2.45 billion monthly active users. Facebook is now the biggest social network.[126]

Evangelization has been a mission of the Church from the beginning. Jesus said to his apostles, "Go, therefore, and make disciples of all nations, baptizing them in the name of the Father, and the Son, and the Holy Spirit, teaching them to observe all that I have commanded you. And behold, I am with you always, until the end of the age" (Matthew 28:19-20). During his ministry, Jesus came to people, taught and converted them in different places. Jesus had entrusted the

---

[124] Dominick, R.J. *Thy Dynamics of Mass Communication, Media in Transition* (New York: McGraw Hill, 2011), 25.

[125] Meredith Gould, *The Social Media Gospel: Sharing the Good News in New Ways* (Minnesota: Liturgical Press, 2015), 3.

[126] J. Clement, "Facebook: number of monthly active users worldwide," November 19, 2019, accessed December 19, 2019, https://www.statista.com/statistics/264810/number-of-monthly-active-facebook-users-worldwide/

mission to his apostles before he went to heaven. The Church continues this mission to spread God's Word to all people wherever they are. Nowadays, billions of people are using social media. Therefore, the Church sees social media as a great means for sharing the Word of God to them.

Since the beginning, the Church has been using many ways and techniques to reach out to people. Over the course of centuries, thousands of missionaries went to many countries all over the world and taught the Gospel of Jesus Christ to unbelievers. The development of technology gives the Church more means to continue the goal of evangelization. Perhaps, the Church has paid special attention to social communication since the Second Vatican Council. Daniella Zsupan-Jerome states,

> The Second Vatican Council's status as a watershed moment for the Church in the twentieth century has been treated extensively in scholarly literature. Key words to describe the spirit of the council include renewal, reform, aggiornamento (updating), openness, dialogue, and reading the signs of the times. It is significant that the Church enter into conversation regarding the topic of social communication in this spirit. Wisely recognizing the cultural shifts occurring around this topic, the Church began a conversation, seeking to find wisdom moving forward as these new media continued to shape the world.[127]

Social media have developed quickly with new technologies and new ways of communication. And many people

---

[127] Daniella Zsupan-Jerome, *Connected toward Communion: The Church and Social Communication in the Digital Age* (Minnesota: Liturgical Press, 2014), 20.

have participated in social communication. The Church also reaches out to people in social communication and aims to provide social media guidance.

The first two documents published in the Second Vatican Council are *Inter Mirifica* and *Sacrosanctum Concilium*. *Inter Mirifica* is seen as the first decree of the Church in its concern about social media and communication. It has a strong influence on social media in the Church. The first section of *Inter Mirifica* states,

> Among the wonderful technological discoveries which men of talent, especially in the present era, have made with God's help, the Church welcomes and promotes with special interest those which have a most direct relation to men's minds and which have uncovered new avenues of communicating most readily news, views and teachings of every sort. The most important of these inventions are those media which, such as the press, movies, radio, television and the like, can, of their very nature, reach and influence, not only individuals, but the very masses and the whole of human society, and thus can rightly be called the media of social communication.[128]

In this statement of *Inter Mirifica*, the Church asserts that man, with the help of God, invented all inventions of technology regarding social media. In this sense, we may say that all inventions of social media like movies, the radio, televisions... are gifts from God to serve human life. The

---

[128] Paul VI, "Decree on the Media of Social Communications *Inter Mirifica*." December 4, 1963 accessed Dec 19, 2019, https://www.vatican.va/archive/hist_councils/ii_vatican_council /documents/vat-ii_decree_19631204_inter-mirifica_en.html

Church also shows her interest in social communication. Social media and communication could be very helpful in the pastoral ministry of the Church.

In addition, the Church considered media as a good tool to spread the Good News of God to the world, to bring man back to God to have salvation. *Inter Mirifica* expresses, "The Catholic Church since it was founded by Christ our Lord to bear salvation to all men and thus is obliged to preach the Gospel, considers it one of its duties to announce the Good News of salvation also with the help of the media of social communication and to instruct men in their proper use."[129] In terms of Shaun McAfee, *Inter Mirifica* is the foundation for later teaching on social media. He writes, "*Inter Mirifica* is the mortar of all media-related teachings of the Church because, as an ecumenical decree, it is the document from which other teachings have emanated."[130]

Pope Benedict XVI on the 44th World Communications Day emphasized the important pastoral area of social communications in the digital world. He reaffirmed how the Church has been using social media to promote common goods for humanity. He writes, "The Church communities have always used the modern media for fostering communication, engagement with society, and, increasingly, for encouraging dialogue at a wider level. Yet the recent, explosive growth and greater social impact of these media make them all the more important for fruitful priestly ministry."[131] The

---

[129] Ibid.

[130] Shaun A. McAfee, *Social Media Magisterium: A No-Nonsense Guide to the Proper Use of Media* (St. Louis, Missouri: En Route Books and Media, LLC, 2018), 12.

[131] Benedict XVI, "Message of his Holiness Pope Benedict XVI for the 44th World Communications Day: The Priest and Pastoral Ministry in a Digital World, New Media at the Service of the World." May 16, 2010 accessed Dec 19, 2010, http://w2.vatican.va/

Church sees the importance and convenience of social media. Therefore, the Church wants to use it to spread God's word and to promote a good relationship in our society.

Also, Pope Benedict reminds priests of the expectation to involve in social communication with others in the role of leaders. They may use all the means that social media provides for evangelization. Pope Benedict states,

> Yet priests can rightly be expected to be present in the world of digital communications as faithful witnesses to the Gospel, exercising their proper role as leaders of communities which increasingly express themselves with the different "voices" provided by the digital marketplace. Priests are thus challenged to proclaim the Gospel by employing the latest generation of audio-visual resources (images, videos, animated features, blogs, websites) which, alongside traditional means, can open up broad new vista for dialogue, evangelization and catechesis.[132]

Priests and other leaders of the Church can create accounts on social media such as Facebook, Twitter, YouTube, Google... so that they can connect and be friends with many parishioners in parishes, or even people outside the Catholic Church. Through such social media, priests can share daily readings, Gospels, homilies. They even can post information about upcoming events, pictures and videos of other activities like the First Communion and Confirmation classes and Bible study class in their parishes.

---

content/benedict-xvi/en/messages/communications/documents/
hf_ben-xvi_mes_20100124_44th-world-communications-
day.html

[132] Ibid.

Another time, Pope Benedict XVI in the 47th World Communication day in 2013 confirmed the advantage of using social networks to build and foster the human relationship. Social media create a platform in which people can interchange information, ideas and opinions. This way may form new relationships and communities. The Pope explains,

> The development of social networks calls for commitment: people are engaged in building relationships and making friends, in looking for answers to their questions and being entertained, but also in finding intellectual stimulation and sharing knowledge and know-how. The networks are increasingly becoming part of the very fabric of society, inasmuch as they bring people together on the basis of these fundamental needs. Social networks are thus nourished by aspirations rooted in the human heart.[133]

Along with Pope Benedict, Pope Francis also expressed his thought on social media. For him, the development of technology brings many people closer to one another in the sense that they can communicate easily and immediately. Physical distance is not as big problem as it used to be in the past. By the advantage of communication in social media, people can promote solidarity in their societies. Pope Francis describes,

---

[133] Benedict XVI, "Message of his Holiness Pope Benedict XVI for the 47th World Communications Day: Social Networks, portals of truth and faith; new spaces for evangelization." May 12, 2013 accessed December 19, 2019, http://www.vatican.va/content/benedict-xvi/en/messages/communications/documents/hf_ben-xvi_mes_20130124_47th-world-communications-day.html

In a world like this media can help us to feel closer to one another, creating a sense of unity of the human family, which can in turn inspire solidarity and serious efforts to ensure a more dignified life for all. Good communication helps us to grow closer, to know one another better and ultimately, to grow in unity.[134]

For Pope Francis, communication truly draws us closer to others. It is by communication that we become neighbors and friends. He used the story of the Good Samaritan as an example of the effect of communication. In the story, there was a man who was beaten by robbers, and he was left abandoned on the road. The Levite and the priest, who have believed in God, ignored the man. They saw the man but turned to another road. The Good Samaritan did not know the man. But when he saw the man in half-dead condition, the Good Samaritan took care of him. The central question now is that what makes us neighbors? Pope Francis explains, "It is not just about seeing the other as someone like myself, but of the ability to make myself like the other."[135] Therefore, neighbors are not just seen as people who live near us in society, but communication truly makes us become neighbors. Neighbors without communication maybe not true neighbors. Social communication helps us to know ourselves better in relation with God and with others. Pope Francis states, "Communication is really about realizing that we are all human beings, children of God. I like seeing this power of

---

[134] Francis, "Message of Pope Francis for the 48th World Communications Day: Communication at the Service of an Authentic Culture of Encounter." June 1, 2014 accessed Dec 19, 2019, https://fwdioc.org/pope-francis-world-communications-day-6-1-14.pdf

[135] Ibid.

communication as 'neighborliness'."[136]

Nevertheless, there are challenges and problems in the current state of social media. Pope Benedict XVI shows the first challenge is how to determine and speak the truth and values in social media. He says,

> The culture of social networks and the changes in the means and styles of communication pose demanding challenges to those who want to speak about truth and values. Often, as is also the case with other means of social communication, the significance and effectiveness of the various forms of expression appear to be determined more by their popularity than by their intrinsic importance and value. Popularity, for its part, is often linked to celebrity or to strategies of persuasion rather than to the logic or argumentation.[137]

According to Pope Benedict, social media and those who participate in media must commit themselves to the value of dialogue, logical argumentation and reasoned debate. In this way, dialogue and discussion on social media can gather people together in unity because people respect the truth and respect different opinions from others. The Gospel message and values of human dignity can be shared broadly in social media.

Pope Francis also shows other problems in social media. The problem is that we sometimes do not know how to deal

[136] Francis, "Message of Pope Francis for the 48th World Communications Day: Communication at the Service of an Authentic Culture of Encounter."

[137] Benedict XVI, "Message of his Holiness Pope Benedict XVI for the 47th World Communications Day: Social Networks, portals of truth and faith; new spaces for evangelization."

with different information, opinions and thoughts. They may make us confused about how to make the right judgments. The pope explains,

> The speed with which information is communicated exceeds our capacity for reflection and judgement, and this does not make for more balanced and proper forms of self-expression. The variety of opinions being aired can be seen as helpful, but it also enables people to barricade themselves behind sources of information which only confirm their own wishes and ideas, or political and economic interests. The world of communication can help us either expand our knowledge or to lose our bearings.[138]

In reality, we can see the division in politics. Many people use social media not to communicate with others, not to share a good message of truth and love. Rather, they use social media to divide others. They do not accurately report news and events. They distort the truth or create fake news because they try to convince others to agree with their opinion.

Social media is a great tool for us to get information. We can use it for evangelization and building good relationships. Fr. Jay Finelli asserts, "At the same time, the use of social media is not only for getting our message out and sharing information, but it's also about building relationships with real people who create bonds with one another, and this opens hearts to what it is we have to share."[139] In using social media

---

[138] Francis, "Message of Pope Francis for the 48th World Communications Day: Communication at the Service of an Authentic Culture of Encounter."

[139] Jay Finelli, "The New New Evangelization: Embrace sharing the faith on social media," Vol. 73 Issue 8 (August 2017): 33.

for this purpose, evangelizers and participants in social media need to follow the rules and guidelines. They must respect boundaries for communication. For priests and other leaders of the Church, they need to follow codes of conduct of the diocese and communities for posts and comments on social media. Furthermore, evangelizers in the media should provide their way and recommendations on how to deal with difficulties.

After all is set up, I have to evaluate the effectiveness of Facebook for my purpose in promotions Bible study class and good relationship among parishioners in St. Cecilia's Church. In order to evaluate how successful my own outreach is, I will check how many people are connected to my Facebook account. Within a short period, I will check the number of visitors, numbers of page views, new members. It is very important to evaluate the quantity and quality of comments, posts, updates, reflections on Scriptures and spiritual thoughts. For better evaluation and good development in the future, multiple metrics could be used to have an accurate assessment.

7

Our Christian Faith between Normal Preaching and Facebook Preaching

George Ziadeh, Seminarian,
Latin Patriarchate of Jerusalem

Introduction

People are connecting now more than ever through social media and the World Wide Web. Over than 270 million people in North America use the Internet, and that number is rapidly increasing, and about 73% of the US users of the Internet are reached by *Facebook*.[140] Social media catches the eye of the world if used correctly, and many Catholic groups are doing just that. We can find in our days that a lot of catholic people and others do pages, events, share pictures, videos, quotes, etc... on Facebook and on other social media platforms. The potential for evangelization through this medium is almost limitless and many are yet to explore all its potential. Through social media, the Roman Catholic Church has the potential to

---

[140] Nick Rabiipour, "The Importance of Social Media on Evangelization." (September 1, 2011). https://www.getfed.com/importance-social-media-evangelization-5521/

promote the Gospel by promoting Catholic books, the rosary and more.

During this era—the era of rapid technological development and progress—we must use all means available to reach the largest number of people, where we can preach them, teach them and educate them religiously. Talking about the evangelization and the importance of giving this word of God to all the nations, St. John says: "what we have seen and heard, we proclaim now to you, so that you too may have fellowship with us; for our fellowship is with the Father and with his Son, Jesus Christ" (1 John 1:3).[141] The Lord said to Paul, "Do not be afraid. Go on speaking, and do not be silent, for I am with you" (Acts 18:9-10a).[142] If the Lord says to St. Paul to preach and he will be with him, what keeps us afraid of using social media to preach and give the faith to our people?

Facebook Page

<u>https://www.facebook.com/My-Christian-FAITH-112466033451174/</u>

This link is about my Facebook page which is called: "My Christian FAITH". This page was created during a course entitled Social Media: Introduction to Theology and Practice that was offered by Dr. Sebastian Mahfood, OP, at St. Joseph's Seminary-Dunwoodie in Yonkers, NY, at the beginning of the 2019-2020 academic year.

The page is a comprehensive page: it includes many important matters to the Christian life and works to help develop the Christian concept of private and public faith

---

[141] http://www.usccb.org/bible/1john/1
[142] http://www.usccb.org/bible/acts/18

through sharing pictures, videos, spiritual ideas, thinking from saints, news, saints' feasts, an idea from the saint of that day, quotes...etc. It actually helps develop a broader and deeper understanding of the Christian faith in various aspects, such as morality, the Bible and its interpretation, doctrinal aspects... and so on.

I chose this picture which follows because I would like to make a link between the page on Facebook, which I created, and the name of this page, which is faith. This image explains everything I want to do on this page: Faithbook. For sure we are not familiar with the name, and maybe this is the first time that we have heard or seen it, but this really explains to us that it is possible to share our faith on Facebook. It's doubtless that its success will be quick because a lot of people use social media daily, and such use will give them an opportunity to find their faith.

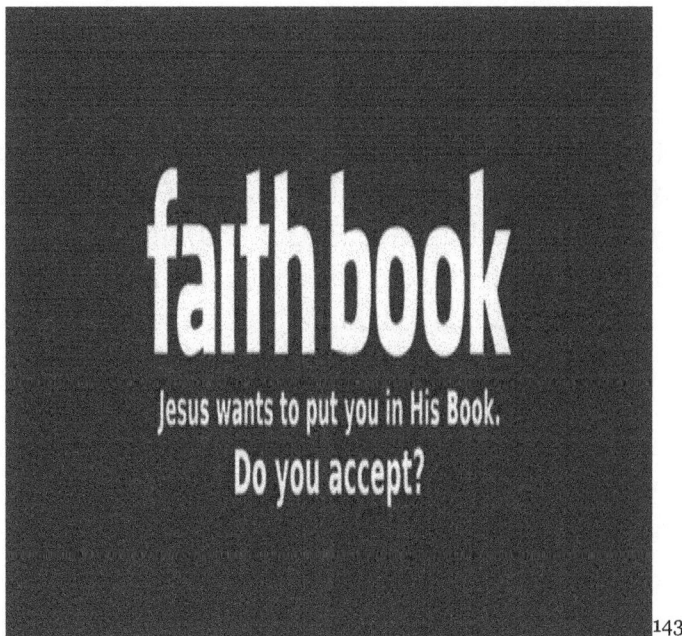

[143]

---

[143] https://fbcover.com/facebook/religious/faithbook.html

There is a question under the title that asks, "Jesus wants to put you in His Book. Do you accept?" This is an amazing question, which we must ask ourselves about. Do we really want to enter this relationship with God, especially when the opportunity in front of us is through the social media?

Church Teaching - Evangelization in Social Media

What we have to understand before the explanation of the Church teaching is that the Catholic Church does not change from time to time in its teachings, but it does find new methods to preach to the current age and uses the resources available to people—not just for the Church itself—to enable the proclamation of the Gospel of the Lord.

The Church always talks about social media and its importance for our preaching. In the past few years, Catholics and non-Catholic Christians have been actively engaging the different forms of social media platforms with the purpose of spreading the Gospel and catechizing. We can see as an example the Archbishop of Jerusalem–His Excellency Pierbattista Pizzaballa—who shares his homilies and meditations every week on a private Facebook page in three different languages. Also, His Eminence Cardinal Dolan shares a video from time to time on his own page on Facebook, talking about different things, saints, spiritual ideas... and so on. That for sure gives us a power to teach and preach with these things that people use. The Church's interest in the Internet is an expression of her longstanding interest in the media of social communication, seeing the media as an outcome of the historical scientific process of humankind.[144]

---

[144] Pontifical Council for Social Communications, "The Church and Internet." (February 22, 2002). http://www.vatican.va/roman

Pope Pius XII's *Miranda Prorsus* (1957)

Pope Pius XII in his encyclical letter *Miranda Prorsus* says, "The Church sees these media as 'gifts of God' which, in accordance with his providential design, unite men in brotherhood and so help them to cooperate with his plan for their salvation".[145] As a gift of God, the media provide us with a formula: our following his plan by using his gifts will lead to our salvation. This is amazing. Pope Pius XII elevated social media to the level of God's gifts for our salvation, not because of the media alone but because of the means they provide for us to proclaim the Good News about our salvation.

St. Pope Paul VI's *Inter Mirifica* (1963)

In his 1963 Decree on the media of social communications *Inter Mirifica,* promulgated through the Second Vatican Council in 1963,[146] St. Pope Paul VI wrote that the Church recognizes that these media, if properly utilized, can be of great service to mankind. He focused on two points for evangelization on media: Pastors should hasten to fulfill their duty in this respect, one which is intimately linked with their ordinary preaching responsibility, and the laity, too, who have something to do with the use of these media, should endeavor to bear witness to Christ through them.

_curia/pontifical_councils/pccs/documents/rc_pc_pccs_doc_20 020228_church-internet_en.html

[145] Pius XII, *Miranda Prorsus* (September 8, 1957). Online at http://www.vatican.va/content/pius-xii/en/encyclicals/ documents/hf_p-xii_enc_08091957_miranda-prorsus.html

[146] Paul VI, *Inter Mirifica*, Second Vatican Council (December 4, 1963), online at http://www.vatican.va/archive/hist_councils/ ii_vatican_council/documents/vat-ii_decree_19631204_inter-mirifica_en.html

George Ziadeh

St. John Paul II's Many Promulgations

In 2002, Pope St. John Paul II invited the Church to "put out to sea" in evangelizing the world with the help of the internet.[147] For him, media play the key role in the work of evangelization. In his message for the 34th World Communications Day,[148] for instance, John Paul II started with this verse from the book of Acts, which reads, "You shall be my witnesses in Jerusalem and in all Judea and Samaria and to the end of the earth" (Acts 1:8). That means from the beginning of the Church, Jesus asked his apostles to be His witnesses to the ends of the earth. He continues, in proclaiming the Lord, the Church must make energetic and skillful use of her own means of communication, including books, newspapers and periodicals, radio, television and other means. Catholic communicators must be bold and creative in developing new media and methods of proclamation. But, as much as possible, the Church also must use the opportunities that are to be found in the secular media.[149]

PCSC's "The Church and Internet" (2002)

The Pontifical Council for Social Communications, a dicastery formed out of Vatican II and merged into the Dicastery for Communications in March of 2016, focuses in its 2002 document "The Church and Internet" on the following

---

[147] Michael, "What John Paul II said about the Internet." (December 29, 2016) https://www.michaeljournal.org/articles/other-topics/item/what-john-paul-ii-said-about-the-internet

[148] John Paul II, "Proclaiming Christ in the Media at the Dawn of the New Millennium." (June 4, 2000). http://www.vatican.va/content/john-paul-ii/en/messages/communications/documents/hf_jp-ii_mes_20000124_world-communications-day.html

[149]

points:

1. The Church need to understand the media and apply this understanding in formulating pastoral plans for social communications.
2. The Church leaders' obligation to use "the full potential of the 'computer age' to serve the human and tran-scendent vocation of every person, and thus to give glory to the Father from whom all good things come."[150]
3. The urgent need to evangelize in the age of the Internet, with its enormous outreach and impact.
4. The need for Church reflection on the Internet, as upon all the other media of social communications, in consideration of Christ as "the perfect communi-cator."[151]
5. The hope that Catholics involved in the world of social communications preach the truth of Jesus evermore.

Pope Benedict XVI

Pope Benedict XVI: "Young people in particular, I appeal to you: bear witness to your faith through the digital world!"[152] In 2010, during his meeting with priests, he urged them to use the Internet to spread the word of God.[153]

---

[150] Message for the 24th World Communications Day, 1990.

[151] *Communio et Progressio*, n. 11.

[152] Catholic News Agency, "Pope Calls for Internet Evangelists" (May 20, 2009) https://www.catholicnewsagency.com/news/ pope_calls_ for_internet_evangelists

[153] VOA News, "Pope Encourages Catholic Priests to Use Internet" (January 22, 2010) https://www.voanews.com/europe/ pope-benedict-encourages-catholic-priests-use-internet

Pope Francis

Pope Francis in "Communication and Mercy: A Fruitful Encounter," for the 50th World Communications Day, wrote that "[t]ext messages and social media are a gift from God" and that "[e]mails, text messages, social networks and chats, can be fully human forms of communication."[154]

Conclusion

I believe that a lot of people would follow my page to know more because I would share interesting things that will give them a better understanding about Christ. Many people would be interested in videos that would provide them with a better understanding and a good knowledge of their Christian faith. I would measure my impact by seeing how many subscribers reconsider their faith, how many can spread their faith, and how many can maintain their faith in these difficult circumstances. It would also be useful to establish a correlation between subscription to my page and an increase in mass attendance.

At the end, we must remember the good work that has been done by the Church and what the Church gave and still gives us during our days. I hope that Christians will be able to use this page and pages like it that can help them to go more deeply into their faith and see that God is with them always. All popes over the past 60 years have tried to advance the work of the Church through these huge developments in communication in order to make a new kind of evangelization, with

---

[154] *Catholic Herald*, "Pope Francis: 'Text Messages and social media are a gift from God'" (January 25, 2016) https://catholicherald.co.uk/news/2016/01/25/pope-francis-text-messages-and-social-media-are-a-gift-from-god/

things that people use and have. The impact of social media platforms is that they allow us to reach a lot of people at the same time and provide a lot of materials for deepening their faith, providing a strong Christian life, for developing a greater love of Christ and his Church.

Part II

Aquinas Institute of Theology and Holy Apostles College & Seminary

Two Selections

8

Social Media and Rural Evangelization: A Model for the New Evangelization

Fr. Dominic Ibok
Diocese of Springfield-Cape Girardeau, MO
Aquinas Institute of Theology, St. Louis, MO
D.Min. Program

Social media have provided an important means of communication, and with the ongoing developments and innovations in this area, the Church has seen its usefulness in the work of promoting the Good News. Our Holy Father, Saint John Paul II, was not hesitant in noting that

> The advent of the information society is a real cultural revolution, making the media "the first Areopagus of the modern age" (_Redemptoris Missio_, 37), where facts and ideas and values are constantly being exchanged. Through the media, people come into contact with other people and events, and form their opinions about the world they live in - indeed, form their understanding of the meaning of life. For many, the experience of living is to a great extent an experience of

the media (cf. Pontifical Council for Social Communications, *Aetatis Novae*, 2). The proclamation of Christ must be part of this experience.[155]

As one who ministers in a rural community, I have seen how the use of social media have affected my community. The words of our Holy Father, Saint John Paul II, therefore, set the stage for this paper as I delve into how social media have shaped my work in rural Southeast Missouri, specifically in the area of preaching and evangelization. One of the areas is that of networking. While this could certainly have a variety of meanings, for this paper, I will describe networking within my ministerial context as a way of dialogue, interaction and connectivity among members within the community including both Catholics and non-Catholics. Many who live within these communities reside on vast acreages of property. Location and distance in a rural community in many ways could be a limiting factor when it comes to evangelization. This, therefore, limits a frequent personal and face-to-face engagement, not to mention the seasonal work of farming, which has its unique challenge. It is in this context that social media have become a viable tool to bridge the gap of communication and socialization, albeit on a different level of interaction. The documents from the Church—before, during, and after the council—on social media and its use have contributed to inviting everyone, especially as it pertains to evangelization, to a renewed engagement of evangelization in new forms and methods. Social media have become not only

---

[155] John Paul II, *Proclaiming Christ in the Media at the Dawn of the New Millennium,* accessed online August 6, 2019 at http://w2.vatican.va/content/john-paul-ii/en/messages/communications/documents/hf_jp-ii_mes_20000124_world-communications-day.html.

a tool but also a channel that presents clergy and laity within the Church with an opportunity to work in the field of evangelization. Pertinent to this work is *Inter Mirifica*, the *Decree on Social Communications* issued during the council. This document provides an invitation to all the faithful—clergy and laity—on the need for promoting the work and mission of the Church through social media.

This research paper will attempt to propose how rural evangelization can benefit from the use of social media. While a considerable number of articles and materials promote the use of social media, these have often focused on urban areas and cultures. In writing this paper, I realize that some rural communities lack the opportunities present in urban areas like easy access to personnel and materials for evangelization, yet this should not prevent our engaging the people who reside in these areas but see it as an opportunity to promote the Good News in new and different ways. The decree notes that

> The Catholic Church, since it was founded by Christ our Lord to bear salvation to all men and thus is obliged to preach the Gospel, considers it one of its duties to announce the Good News of salvation also with the help of the media of social communication and to instruct men in their proper use. It is, therefore, an inherent right of the Church to have at its disposal and to employ any of these media insofar as they are necessary or useful for the instruction of Christians and all its efforts for the welfare of souls. It is the duty of Pastors to instruct and guide the faithful so that they, with the help of these same media, may further the salvation and perfection of themselves and of the entire human family. In addition, the laity especially must strive to instill a human and Christian spirit into these

media, so that they may fully measure up to the great expectations of mankind and to God's design.[156]

The council does not hesitate to note that while we are under obligation to preach, the media of social communication provides an opportunity for us to accomplish this aspect of our mission. The Church teaching not only gives us a foundation for engagement but also notes how social media engagements can provide an opportunity for Catholics—active and inactive—to grow in their appreciation of the faith and to provide a means for greater participation in the sacraments. The decree also encourages all the faithful to acquire the knowledge and skills that will make this mission possible. Paragraph 13 states that

All the children of the Church should join, without delay and with the greatest effort in a common work to make effective use of the media of social communication in various apostolic endeavors, as circumstances and conditions demand. They should anticipate harmful developments, especially in regions where more urgent efforts to advance morality and religion are needed. Pastors should hasten, therefore, to fulfill their duty in this respect, one which is intimately linked with their ordinary preaching responsibility. The laity, too, who have something to do with the use of these media, should endeavor to bear witness to Christ, first of all by carrying out their

---

[156]Vatican II, *Decree on the Media of Social Communications*, par. 3, accessed online August 6, 2019 at http://www.vatican.va/roman_curia/pontifical_councils/pccs/documents/rc_pc_pccs_doc_04121963_inter-mirifica_en.html.

individual duties or office expertly and with an apostolic spirit, and, further, by being of direct help in the pastoral activity of the Church—to the best of their ability—through their technical, economic, cultural and artistic talents.[157]

This aspect of the exhortation invites all members of the Church—clergy and laity—to reflect on the importance of social media and how our engagement in it can facilitate our work of catechesis and evangelization. Collaboration with the laity who are well-versed in the use of social media and its operation is also necessary for arriving at a common goal.

Consequently, the theology of social media highlights a variety of areas where social media can aid evangelization and how that could translate in the work of rural evangelization. Before its suppression in 2016, the Pontifical Council for Social Communications provided a few pointers that could help as a guide for the use of social media. Saint Paul VI in his message for World Communications Day notes that "Evangelization is an integral part of the mission of the Church, sent by Christ into the world to preach the Gospel to every creature (*Mk* 16,15). The Church carries out this duty above all in her liturgical life, but she is constrained to fulfill it also in all the ways and by all the means which are available to her to use as she abides among the people of each continent."[158]

---

[157] Ibid., par. 13.

[158] Paul VI, *Social Communications and Evangelization in Today's World,* accessed online August 6, 2019 at http://w2.vatican.va/content/paul-vi/en/messages/ communications/documents/hf_p-vi_mes_19740516_viii-com-day.html.

In looking at the past fifty years of writings and engagement between the Church and the media world, one cannot help but observe how a lot has changed in both the secular and ecclesiastical circles. While it may seem like the Church has moved with baby steps, the secular world continues to engage the world of social media with giant steps. One could argue, moreover, that there are both advantages and disadvantages to these movements. These developments have affected the Church in the twenty-first century and particularly rural evangelization today.

Why Use Social Media?

The groundwork highlighted by the decree on Social Communications presents us with a strong basis for action. After Vatican Council II, the use of social media slowly evolved within the Catholic Church as a means of outreach and evangelization. Nevertheless, this progress was limited and restrained until 1971 when Saint Paul VI presented the entire Church with a fresh challenge noting that

> The modern media offer new ways of confronting people with the message of the Gospel, of allowing Christians even when they are far away to share in sacred rites, worship and ecclesiastical functions. In this way they can bind the Christian community closer together and invite everyone to participate in the intimate life of the Church. Of course the mode of presentation has to suit the special nature of the medium being used. The media are not the same as a Church pulpit. It cannot be overstressed that the

standard of such presentations must at least equal in quality the other productions of the media.[159]

Modern media offer us new methods of promoting the message of the Gospel, and the ways these methods aid us in our evangelizing efforts are important to note. In many parts of the world, the media are reaching people and bringing information to homes in ways that can either be promising or endangering. It is up to us as modern-day missionary disciples to ensure that the media are used to provide important channels for evangelization and catechesis.

In an age during which the world is constantly becoming smaller, becoming *a global village*, it is no surprise that social media have become a powerful means by which to interact and connect with people both locally and internationally. However, one must look at how social media can benefit the Church's mission at multiple levels as the Holy Father, Saint Paul VI, notes

The media are invaluable helps for Christian education. They can call on the services of the greatest specialists in religious teaching as well as of experts on all the questions that arise. The media have at their command all the technical facilities required for attractive and contemporary presentation. They can back up most effectively the personal work of the daily teacher. Their resources make possible the radical changes that are required in the whole style of religious

---

[159] Pope Paul VI, *The Role of Communications Media in Promoting Unity Among Men*, accessed online August 6, 2019 at http://w2.vatican.va/content/paul-vi/en/messages/communications/documents/hf_p-vi_mes_19710325_v-com-day.html, par. 128.

instruction today. Since the instruments of social communication are the usual channels for giving the news and voicing contemporary man's attitudes and views, they offer marvelous opportunities to all for considering the deeper implications of their religious convictions through the discussion of events and problems of the day. The Christian can then apply these deepened convictions to his daily life.[160]

The value of acquiring and applying social media as part of our mission can, therefore, enhance our evangelizing efforts in ways that exceed our traditional methods of communicating. With easy access today, social media seem to be the way to go in developing a new method of evangelization. Its newness lies in applying the content of our message through a contemporary channel to access a greater audience beyond the parish boundaries or community.

In my ministerial setting in Pemiscot and New Madrid counties, as it is observable in other areas, both young and old are constantly engaging each other through social media tools like Facebook, Twitter, Instagram and others. A Catholic presence on these web-based tools is an opportunity to enter their world and to be a presence that offers hope and alternatives that would, in some cases, be life-giving to their search for answers. Marcel LeJeune notes in an article in the edited work by Brandon Vogt that "we must reach out, engage, and evangelize our young adults or lose them to the secular culture or other non-Catholic groups."[161] In making this comment, LeJeune highlights the evangelization of the youth, which is key to the future of our parishes.

---

[160] Ibid., par. 129.

[161] Brandon Vogt, *The Church and the New Media*, (Indiana: Our Sunday Visitor, 2011), 59.

As a preacher, social media facilitates a viable web presence. The experience of parish ministry today requires a social media presence considering the many conflicting voices in the media that vie for the attention of our members. In my ministry, I have developed a Facebook page and an internet radio link in collaboration with WCAT radio where my homilies can either be viewed or listened.[162] The parish community is one area where the community is built and renewed, but the virtual community reached through social media is another area where the Word engaged in the liturgical space is perpetuated throughout cyberspace. The virtual community, it should be noted, is comprised of real communities though engagements within any given virtual community may be limited to the task at hand.

Both social media tools I have employed, namely internet radio and Facebook, have yielded some results which, as I continue to review the data, have made me more cognizant of the needs of my audience. The message for world communications day from Saint John Paul II reminds us that

> Since at the beginning of communication there is a man-communicator and, at its end, there is a man-receiver, the media of social communication will facilitate the encounter between faith and culture the more they foster the encounter of persons among themselves, so that a mass of isolated individuals will not be formed, each one of whom may be in dialogue with the written page, or the stage, or the small or large screen, but a community of persons aware of the importance of the meeting with faith and with culture and determined to achieve it through personal contact,

---

[162] See https://www.wcatradio.com/fromtheambo

in the family, in the place of employment, in social relations.[163]

As a preacher and minister of the Gospel, I find that my leaning into the benefits of communicative media as a way of reaching out to members of my parish community and the secular society is a step towards bridging the gap between faith and cultures. How so? When one engages in social media and adopts them as means of proclaiming the Gospel message, one finds that they open up opportunities for ongoing dialogue and interaction within the community. Noteworthy is the fact that

> If Catholics are to be of service to the means of social communication and to act so that these may serve humanity's ends, it goes without saying that it is in the spiritual sphere that the Church can best help. The Church hopes that, as a result of her spiritual contribution, the basic nature of social communication will be more clearly appreciated. The Church hopes, too, that the dignity of the human person, both communicator and recipient, will be better understood and respected. In this way this social interplay that makes neighbors of men can lead to true communion.[164]

Therefore, part of the service we offer to the world is reaching out through social media to provide those who are

---

[163] John Paul II, *Social Communication: Instruments of Encounter Between Faith and Culture,* accessed online August 6, 2019 at http://w2.vatican.va/content/john-paul-ii/en/messages/communications/documents/hf_jp-ii_mes_24051984_world-communications-day.html, par. 6.

[164] Paul VI, *The Role of Communications Media,* par. 102.

still searching an opportunity to engage the Good News while being a presence to them in cyberspace. Saint John Paul II reminds all the faithful that

> The Internet causes billions of images to appear on millions of computer monitors around the planet. From this galaxy of sight and sound will the face of Christ emerge and the voice of Christ be heard? For it is only when his face is seen and his voice heard that the world will know the glad tidings of our redemption. This is the purpose of evangelization. And this is what will make the Internet a genuinely human space, for if there is no room for Christ, there is no room for man.[165]

The internet has become a thriving platform for every kind of communication. As evangelizers, we are invited to see this opportunity as one that allows us to reach out to a unique audience. However, it would be difficult not to notice the far-reaching possibilities which the use of social media can bring to a parish community. One area that comes to mind is that of catechesis. In a rural farm community, one of the challenges we face is how to continue our dialogue and engagement after the Sunday liturgy. While a few parishioners can attend classes offered within the parish, my observation is that most of the parishioners are Facebook users; hence, a presence on social media. Facebook group messages that facilitate our virtual dialogue to strengthen the faith of our members has developed into an important aspect of our connectivity and

---

[165] John Paul II, *Internet: A New Forum for Proclaiming the Gospel,* accessed online August 6, 2019 at http//w2.vatican.va/content/john-paul-ii/en/messages/communications/documents/hf_jp-ii_mes_20000124_world-communications-day.html.

interaction on the web. The message of our Holy Father Saint John Paul II highlights that

> With the advent of computer telecommunications and what are known as computer participation systems, the Church is offered further means for fulfilling her mission. Methods of facilitating communication and dialogue among her own members can strengthen the bonds of unity between them. Immediate access to information makes it possible for her to deepen her dialogue with the contemporary world. In the new "computer culture" the Church can more readily inform the world of her beliefs and explain the reasons for her stance on any given issue or event.[166]

Our rural farm community has benefitted greatly from our online presence and the use of social media. We have seen an increase in participation and the desire of a few who were inactive within the parish community slowly coming back and asking questions that lead to deeper dialogue. This has been a promising result due to the use of social media within our parish community.

The parish website is another area where we extend our parish community's presence. As regards the website, many are informed about activities within the parish community. While this might sound basic, in a rural farm community, parishioners who go on vacation and other trips comment that their connection with the local parish community is made possible through our web presence. Each member of our

---

[166] John Paul II, *The Christian Message in a Computer Culture*, accessed online August 6, 2019 at http://w2.vatican.va/content/john-paul-ii/en/messages/communications/documents/hf_jp-ii_mes_24011990_world-communications-day.html.

parish is assigned a code which grants them access to the Facebook page and allows them to access the feed. Our parish secretary works in collaboration with me to inform the parishioners on the various homilies and information that affect our parish community and anything that needs our communal participation. Cognizant of this, we continue to invest in the development of our website as a means of community networking and outreach. The website has been developed to have links to our parish Facebook page that gives additional information and links to important resources and aids to those who are returning Catholics or those in search of a parish family. These are only a few ways in which social media use continues to promote our work of evangelization and parochial presence within cyberspace.

Magisterial Exhortations

The magisterial documents on social media engagement challenge priests to see how these social media tools could be used in ministerial settings or within the parish. In the past ten years, many documents and messages from our Popes have emphasized the need for engaging through social media. Pope Benedict XVI writes that

> The spread of multimedia communications and its rich "menu of options" might make us think it sufficient simply to be present on the Web, or to see it only as a space to be filled. Yet priests can rightly be expected to be present in the world of digital communications as faithful witnesses to the Gospel, exercising their proper role as leaders of communities which increasingly express themselves with the different "voices" provided by the digital marketplace. Priests are thus challenged to proclaim the Gospel by employing the latest genera-

tion of audiovisual resources (images, videos, animated features, blogs, and websites) which, alongside traditional means, can open up broad new vistas for dialogue, evangelization and catechesis.[167]

This invitation emphasizes how social communication is an important aspect of the propagation of the Gospel. In rural parish ministry, the use of social media tools aids in reaching out especially to inactive Catholics who are logged on online. In Sacred Heart Parish, Caruthersville, which is one of my ministerial sites, we currently have about thirty-six inactive members. Of this number, about twenty are engaged with us online due to our web presence. Seeing such opportunities, we use social media as a way of extending invitations to these members who may desire to return to life within the Parish and an encounter with Jesus in the sacraments.

However, as we strive to use these tools to disseminate the word of God as a way of evangelizing our local parish community, it is important to note an earlier invitation from Pope Pius XII in his encyclical on the use of the radio, television and motion pictures that these means of communication can "be at the service of truth in such a way that the bonds between peoples may become yet closer."[168] In expressing this invitation, the Holy Father highlights how the

---

[167] Benedict XVI, *The Priest and Pastoral Ministry in a Digital World: New Media at the Service of the Word,* accessed online August 6, 2019 at http://w2.vatican.va/content/benedict-xvi/en/messages/communications/documents/hf_ben-xvi_mes_20100124_44th-world-communications-day.html.

[168] Pius XII, *Miranda Prorsus*, Encyclical Letter on the Advancement of Motion Pictures, Radio, and the Television, par. 3, accessed online August 6 at http://w2.vatican.va/content/pius-xii/en/encyclicals/documents/hf_p-xii_enc_08091957_miranda-prorsus.html.

use of social media tools can be a source of community building. The Holy Father's comment also reminds us that the virtual and online community today is equally in need of a Catholic presence that can listen and engage people on various levels. Nevertheless, while we seek to use these tools to our benefit, the caution is that we strive to balance our use of media with our interpersonal relationship with those we are engaging. The Pontifical Council for Social Communication reminds us that "even as the Church takes a positive, sympathetic approach to media, seeking to enter into the culture created by modern communications in order to evangelize effectively, it is necessary at the very same time that the Church offers a critical evaluation of mass media and their impact upon culture."[169] While secular culture promotes the use of social media, we within the Catholic Community need to be prudent and disciplined in our use of and interaction through social media.

Social Media and Preaching

Rural evangelization, as I have experienced, has its unique challenges. The parishioners who are farmers by profession and occupation work at certain times of the year. Depending on the weather, the effects of this seasonal task limits their participation in liturgical celebrations. The use of social media, therefore, acts as a means of keeping them in touch with various activities within the parish community. Social media become an important channel to help them during this

---

[169]Pontifical Council for Social Communications, *Aetatis Novae*, Pastoral Instruction on Social Communication on the Anniversary of *Communio Et Progressio*, par. 12, accessed online August 6, 2019 athttp://www.vatican.va/roman_curia/pontifical_councils/pccs/documents/rc_pc_pccs_doc_22021992_aetatis_en.html.

phase of their Christian journey as Pope Benedict XVI notes when he states that

> Using new communication technologies, priests can introduce people to the life of the Church and help our contemporaries to discover the face of Christ. They will best achieve this aim if they learn, from the time of their formation, how to use these technologies in a competent and appropriate way, shaped by sound theological insights and reflecting a strong priestly spirituality grounded in constant dialogue with the Lord. Yet priests present in the world of digital communications should be less notable for their media savvy than for their priestly heart, their closeness to Christ. This will not only enliven their pastoral outreach, but also will give a "soul" to the fabric of communications that makes up the 'Web.'[170]

While outreach is a considerable need for us in evangelizing, strengthening those within the flock is also important as we use social media. Nevertheless, for preachers, there is a need to be aware of the various areas we can use social media to promote the mission of the Church and the limits of this instrument. Saint John Paul II notes that

> The Church approaches this new medium with realism and confidence. Like other communications media, it is a means, not an end in itself. The Internet can offer magnificent opportunities for evangelization if used with competence and a clear awareness of its strengths and weaknesses. Above all, by providing information

---

[170] Benedict XVI, *The Priest and Pastoral Ministry in a Digital World*, 2010.

and stirring interest it makes possible an initial encounter with the Christian message, especially among the young who increasingly turn to the world of cyberspace as a window on the world. It is important, therefore, that the Christian community think of very practical ways of helping those who first make contact through the Internet to move from the virtual world of cyberspace to the real world of Christian community.[171]

We must not fail to emphasize that social media is a means and while an important means should lead to greater interpersonal conversations among all within and outside the Christian community. The key is remembering that virtue stands in the middle, and we should not move either towards doing too much or doing too little. Our efforts in using social media should be to strive to do what is required in building our faith and the faith of those within our parishes by offering to those who are searching a means to engage the Good News.

Community Formation

Social media presents many opportunities to develop and engage the virtual world. The virtual community is a real presence that needs to be mentioned here because many non-Church goers sometimes see this space as an opportunity to discuss ideas that impact their lives. In recognizing this opportunity for us as Catholics, we see our presence in this virtual community will enable us to offer our ideas and share with those who desire to learn and are open to being formed. The world is gradually becoming a global village as earlier indicated, thanks in part to the use of social media, and our

---

[171] John Paul II, *Internet: A New forum for Proclaiming the Gospel,*2002.

Catholic Faith has one more platform where it can be shared with others especially those outside our parish communities. Nevertheless, while the flattening of the earth and the creation of a global village present us with a lot to expect, one needs to tread with caution.

Furthermore, I should note that the use of new media like Facebook, internet radio, and YouTube has enabled me to expand my reach regarding the preaching of the Word of God. As mentioned earlier, the homily which at one time within my parish community started and ended within the liturgical context now extends to the coffee shop, the pizza hut and the homes of my parishioners thanks in part to our Facebook page and internet radio program. The Farmers Faithbook page which has a link to my YouTube Channel currently has sixty-one friends who are connected. This number is comprised of parishioners from Immaculate Conception Parish, New Madrid and Sacred Heart Parish in Caruthersville. The goal, however, is to reach out to as many parishioners and groups within the area as possible. While this might seem as a long shot, my ongoing development of this project will allow for some creativity and innovation that might facilitate achieving this goal. These media links can be accessed either through smart phones, androids, or the radio. The use of social media have facilitated this engagement, and we are presently training staff and members within the parish community interested in these areas. However, this also allows my parish community to engage each other in the virtual world in ways that were not available before. In line with this development, the words of Saint Paul VI highlight the following:

> Sermons and homilies must be adapted to the nature of the medium that is used. Those who are given the task of preaching in this way, should, therefore, be carefully chosen from among those who have a sound

practical knowledge of the technique of broadcasting. Religious broadcasts, such as newscasts, commentaries, reports and discussions, can contribute a great deal towards education and dialogue.[172]

Through the development of Farmers Faithbook Facebook page—which is located online at https://www.facebook.com/ farmers.faithbook—we find new ways are opening for our parish community as we continue to think of evangelization within our wider community in Caruthersville, Pemiscot County. The various strategies and social media tools utilized have been a beneficial step for our parish family in exploring how we can reach out to the inactive members in our local community. The strategy being used to reach out to the inactive members involves our asking all active members to reach out to them and following up this invitation with a personal visit to the homes of any who are open to a visit.

As regards the weakness of this approach, there is the temptation of our not being able to follow up with some users of this form of media who may have questions. While there may be apps and media tools that can be used to reach out to those who have questions, one will find out that this public forum cannot substitute or replace interpersonal encounters outside of social media. On our Facebook page we do invite all who are friends on the page to make comments or ask questions as the case may be.

Another aspect will be an intentional use of time on this media, which requires discipline and prudence. This means that the Facebook page will be observed during office hours and if there are any questions or comments that need our attention, we will try and contact the persons in need if

---

[172]Paul VI, *The Role of Communications Media,* par. 152 and 153.

possible. My parish secretary at one of our parishes acts as a co-administrator and helps in the video and audio productions. She also tries to get the downloads available as soon as possible and contacts me if there are issues to be discussed. In collaborating with the parish secretary on this, I have an oversight on the page and receive occasional feedback from the parishioners who follow our posts. Regarding training for this, my parish secretary seems to have an advanced knowledge in the area of internet development, and this certainly has been a blessing to our work.

In my ministerial settings, I have had the opportunity to engage both Catholics and non- Catholics, our occasional encounters have been opportunities whereby community and friendships are developed. These have proved to be opportunities for ongoing catechesis and outreach through our interactions on social media. The Decree on Social Communication notes that

> Since the proper use of the media of social communi-cations which are available to audiences of different cultural backgrounds and ages, calls for instruction proper to their needs, programs which are suitable for the purpose—especially where they are designed for young people—should be encouraged, increased in numbers and organized according to Christian moral principles[173]

The inactive Catholics in these rural communities are not limited to the elderly who have suddenly ceased to participate in the life of the Church for various reasons but also the Millennials who do not feel a connection within the Church and many of its practices.

---

[173] Vatican II, *Decree on Social Communications*, par. 16.

One assumption made regarding social media is that it informs and is used widely as an instrument by the young and middle-aged. In my short time of ministry in the *bootheel*, I have found that the Baby Boomers, Gen X, Millennials and Gen Z all engage in the use of social media. An example of how social media ministry has thrived and continues to affect the lives of many in our culture could be seen in Bishop Robert Barron's *Word on Fire* Ministry. My project receives its motivation by the fact that between 80% and 90% of my congregation who are in their 70's have Facebook pages that are active and operative. This select group possesses a Facebook group page that exists as a place for ongoing dialogue and information. The words of the Holy Father Saint Paul VI are pertinent here when he indicates that "Christians must consider how best to employ the instruments of social communication in order to reach countries, societies and persons to whom the apostolate of the Word cannot be brought directly because of particular situations, or scarcity of ministers, or because the Church is unable to exercise her mission freely."[174] One could say that in a rural community where many of the homes are separated by vast acreage of farmland, it seems to me that a way to keep in touch is through social media.

While the usage of social media within this context is unique, it has become a platform for engaging both those within the Catholic community and those outside who have shown a renewed interest in the faith. Within my parish

---

[174] Paul VI, First World Communication Day, *Church and Social Communication,* accessed online August 6, 2019 at http://w2.vatican.va/content /paul-vi/en/messages/communications/documents/hf_p-vi_mes_19670507_i-com-day.html.

communities, about 80% are Facebook users and of this number about 40% are friends on our Facebook page. Since the project is in its early phase, it is our hope to reach out to more parishioners on this project. An ongoing orientation of how this project will benefit our presence and mission in rural Southeast Missouri will also be emphasized.

Because baby boomers, as noted above, are also not hesitant in engaging in the use of social media, we find social media to give us as evangelizers a unique technique to reach out to remote and rural communities and an opportunity for the people to reach us or encounter the message of the Gospel through social communication. History has shown that many missionaries had to deal with diverse cultures, which in some cases were not welcoming of both the missionaries and their messages. Today, with social media as one of our many tools for evangelization in our modern culture and society, we experience a different kind of antagonism which could challenge our communication or propagation of the Good News. If one observes the trend that is developing in our culture today, one will note the following:

We already know that Catholics are no less or more likely than Americans in general to use new technology to access digital or online media. The Church's biggest challenge is to encourage U.S. Catholics to use these devices to connect with their faith more frequently. As the survey indicates, one in five adult Catholics (20 percent) are under the age of 31 and are classified by CARA as members of the Millennial Generation. The oldest members of this generation were in elementary school when the internet began to gain widespread use in the United States. They are sometimes described as the "digital" or "new media" generation. Many assume that the way to connect with this emergent generation of Catholics is not through traditional print media, television, or radio but online—through blogs, Facebook, YouTube, and Twitter. The thought is that if the Church has a presence on these platforms, they will reach these young Catholics. The hope is often stated that we may be able to use new media to get this generation "back into the real world pews" that are more often populated by their parents and grand-parents.[175]

Working with the laity as co-workers and using the means of internet radio and Facebook, the development of a practical way of reaching out to the people in rural Southeast Missouri and possibly other rural communities as a means of evangelization gives us a new opportunity to accomplish our mission of bringing the Good News to all. Looking at my posts on Facebook, I realized that of the sixty-one persons connected to our page; only 16% are regular listeners. The task ahead is getting members of the parish community more engaged so that they could invite others. This is one way we can increase our audience. However, the link of the internet radio needs to be added to our Facebook page as a means of encouraging the parish community in their involvement. Promoting this project in view of our ongoing evangelizing efforts is the next phase of my work. In recent times, Our Holy Father Pope Francis continues to remind the Faithful that

Emails, text messages, social networks and chats can also be fully human forms of communication. It is not

---

[175] Mark M. Gray and Mary L. Gauthier, *Catholic New Media Use in the United States*, accessed online at http://www.usccb.org/about/communications/upload/Catholic_New_Media_Use_in_United_States_2012.pdf, November, 2012.

technology which determines whether or not communication is authentic, but rather the human heart and our capacity to use wisely the means at our disposal. Social networks can facilitate relationships and promote the good of society, but they can also lead to further polarization and division between individuals and groups. The digital world is a public square, a meeting-place where we can either encourage or demean one another, engage in a meaningful discussion or unfair attacks.[176]

The reality of ministry and life today, as disciples called to fulfill our mission of spreading the Gospel message, is that we have a responsibility to engage the culture, we find ourselves whether rural or urban. The rural cultures need not be left out in our evangelizing efforts but included as much as we are able. Nevertheless, my contribution here is to note that in Southern Missouri where the Catholic population is about 5%, our using social media is opening up new opportunities of engagement and evangelization. I hope that with an ongoing outreach and internet radio ministry, the Catholic presence will be strengthened and spread through the use of social media.

---

[176] Pope Francis, *Communication and Mercy: A Fruitful Encounter,* accessed online August 6, 2019 at http://w2.vatican.va/content/francesco/en/messages/communica tions/documents/papa-francesco_20160124_messaggio-comunicazioni-sociali.html.

9

Online Communities:
A Comparative Study of Catholic
Communities on Facebook

Jeremy Chan
Holy Apostles College & Seminary, Cromwell, CT
Master of Arts in Pastoral Studies,
Concentrating in Marriage and Family Studies
Singapore

Introduction

Human community relates to the "personal relationship experienced among individuals that is essential to the constitution of human society"[177] and emerges from the anthropological understanding that "man is constitutionally interpersonal and his inter-relation in community extends the process of the individual's interior self-reflectivity."[178] As more people with shared interest interact on social media,

---

[177] Whitson, R.E.., "Community", in *New Catholic Encyclopedia* (2nd ed, Detroit, MI: Gale, 2013), 38.

[178] "Community", in *New Catholic Encyclopedia*, 39.

regardless of where they are located, these online communities become "real to members who (came) to rely on them for inspiration and support."[179] By online communities, we mean any "virtual space where people come together to converse, exchange information or other resources, learn, play, or just be with each other."[180] This chapter explores online Catholic communities on Facebook in the light of the New Evangelization, i.e., to see how the Church utilizes Facebook as a social media platform as a means of evangelization and building Christian relationships.

Specifically, how do online Catholic communities flourish in terms of their content, interactions and moderation policies? What types of content and production patterns generate the greatest interaction? What kind of infractions occur on the communities and how are they regulated? If we can identify best practices of managing online communities, we can replicate these with others. Online communities can contribute to the Church's mission if we harness the power of new media.

We will evaluate the following four Facebook communities: Catholic Geeks, TOB Institute, Catholics in Singapore! and CatholicNYC.[181] These have been selected due to the author's participation in these groups, as well as to contrast the author's home versus host countries. Also, while TOB Institute represents an institutional community, Catholic Geeks is a group for those with geeky interests.

---

[179] Gould, M., *The Social Media Gospel: Sharing the Good News in New Ways*, (2nd ed, Collegeville, MN: Liturgical Press, St John's Abbey, 2015), 30.

[180] Kraut, E. and Resnick, P., *Building Successful Online Communities: Evidence-based Social Design.* (Cambridge, MA: MIT Press, 2012), 10.

[181] See links to online Catholic communities cited at Appendix A

Research Areas

These online Catholic communities are evaluated against the following yardsticks: (1) content and interaction and (2) moderation policies.

Content and Interaction. Given that communities express the inter-personality of individuals, online communities transform the "Internet from a static source of information into a living, breathing entity."[182] The content and inter-actions on the online communities form the very life of the community as members are "invited to become part of – and indeed to truly shape – the content and the conversation."[183] Instead of a traditional, top-down and producer-to-consumer relationship of media, Facebook offers users a cross-pollination of ideas, content and posts.

In evaluating the Facebook communities on their content and interaction, we will consider Pareto's principle (80/20) and the "90/9/1 Rule (Jakob Nielsen)". Pareto's principle invites social media managers to aim that "80 percent of ... social media content will generate community-building engagement" while limiting broadcasting to "20 percent of ...social media content."[184] The 90/9/1 Rule suggests that 90% are observers and will neither contribute nor participate, 9% will participate occasionally whereas 1% will dominate the community.[185] The latter rule suggests that the reach of the

---

[182] Hendey, L., "That They May Be One: Cultivating Online Community" in Vogt, B., *The Church and New Media: Blogging Converts, Online Activists, and Bishops who Tweet.* (Huntington, IN: Our Sunday Visitor, 2011), 151.

[183] "That They May Be One: Cultivating Online Community" in *The Church and New Media*, 155.

[184] *The Social Media Gospel*, 26.

[185] *The Social Media Gospel*, 26.

online community is more than can be visibly observed on the social media platform.

<u>Moderation</u>. Given the anonymity inherent in social media platforms, the content and interactions on the communities may run contrary to our "longing for living in communion, for belonging to a community"[186] or where communication becomes a form of "violent aggression," primarily aimed at "promoting consumption or manipulating others."[187] To safeguard the authentic culture of encounter in cases of violation, a sound policy of moderation and enforcement is necessary.

It is as though there are two tribes of members on online communities: the Athenians and the Visigoths: "Athenians who follow the rules of civilized online behavior, and Visigoths who break those rules and misbehave."[188] Misbehavior, then, may refer to conduct that does not follow norms, is awry or off. In online communities, it "runs the gamut from playful and roguish pranks, tricks, teasing and mischief... to more serious and malicious offenses, violations, and transgressions."[189] What are the forms of transgressions present in the communities and how are they moderated or disciplined?

---

[186] Pope Francis, "From Social Network Communities to the Human Community," Message for the 53rd World Communications Day, 24 Jan 2019.

[187] Pope Francis, "Communication at the Service of an Authentic Culture of Encounter," Message for the 48th World Communications Day, 1 Jun 2014.

[188] Sternberg, J., *Misbehavior in Cyber Places: The Regulation of Online Conduct in Virtual Communities on the Internet.*, (Lanham, MD: University Press of America, 2012), 12.

[189] *Misbehavior in Cyber Places*, 146.

Content and Interactions

Proving Pareto's principle to be true, the online communities with more promotional/broadcasting content tended to have much fewer interactions amongst the users. A survey of the posts from CatholicNYC from October and November 2019 shows that most of the posts were broadcasting information about events, past and future or live-streams of events.[190] These garnered a few "likes" and other reactions, but there were typically no comments or subsequent interactions from the users.

Nevertheless, CatholicNYC's event invitations appear more professional than Catholics in Singapore! as they utilized Facebook's event feature and often had posters of the event. In the Singapore group, people mainly posted pictures of tickets or print flyers of events and asked if people wanted to go along. There was no RSVP function and the event management seemed haphazard.

The only posts which have gathered more comments are those of digital prayer cards, featuring a saint or an angel, or the meet-the-staff section.[191] While events are frequently uncommented on, digital prayer cards average 3 comments. Their recently launched segment introducing a member of the Office of Young Adult Outreach has received more comments like "What a guy!" or "Absolute Rockstar and Saint!" However, these comments do not reveal much of the users nor do they invite further interaction from other users.

One possible reason accounting for CatholicNYC's mix of content (or rather, a lack thereof) is that the administrators are running the group as a page, i.e., as an alternative to their organization's website. They utilize the event function under

---

[190] Screenshot 1 in Appendix B.
[191] Screenshots 2 and 3 in Appendix B.

Facebook to publicize, obtain RSVPs and to publish photos and videos after each event. Instead of being run like a virtual community, it is mainly treated as a bulletin board and event organization system.

In contrast, the other online Catholic communities surveyed (Catholic Geeks, TOB Institute Community and Catholics in Singapore!) tended to have more varied content, generated by both users and administrators and, consequently, a higher degree of interaction as evidenced by the comments on each post.

One member in the TOB Institute Community reached out to her community for help in preparing a talk on the theology of the body to a small group of men.[192] This generated four suggestions from other users with active replies from the original poster (OP).[193] The OP even provided an update after her talk and shared her slides and audio recording of her lecture. This was then affirmed by another unique user on how important it was for men to be challenged and affirmed by women.

The Conversation Starter feature on Facebook is frequently used in the Catholics in Singapore! group to foster interactions from other users. One member shared a video from South China Morning Post regarding the tragedy of Vietnamese people dying in a refrigerated truck while being smuggled into the United Kingdom.[194] The comments to the original post were generally either outpourings of sympathy, e.g. "It breaks my heart to see or hear all this" or a problem-solving analysis of what accounts for the problem at hand, e.g.

---

[192] Michelle Piccolo, 9 November 2019, 12:17.

[193] Screenshot 4 in Appendix B.

[194] Szeache, 27 November 2019, 22:26.

the promise of "money and (a) better life" being too alluring, leading the victims down a trap.[195]

Furthermore, bearing in mind the 90/9/1 rule which suggests that the proportion of users who are mere observers far outweigh those who actively or even infrequently contribute, the impact of these posts are amplified since the original posts and its subsequent comments may be viewed by other users. For instance, if I were to be invited to give a sharing on the theology of the body to other men but did not want to post within the group, I may browse the comments and content posted by other members. Thus, we would need to consider the knock-on effect of having observers and silent members access the content available on the online communities. Unfortunately, by definition, this is not something that is immediately perceptible on the Facebook communities and would have to be evaluated via either ethnographical methods or survey questionnaires.

While there are more interactions on communities like Catholics in Singapore! and the TOB Institute Community, these interactions tend to be more cerebral rather than vulnerable. Posts were identified as cerebral if they were either debates or did not reveal much personal details of the member. Even in what one would perceive as a more supportive online Catholic community like the one at CatholicMom.com,[196] we find that most of the posts are the sharing of online articles, promotion of books or inspirational memes. While we are not privy to the private messages shared between users nor are we able to observe the offline relationships of the users, this problem may have deeper anthropological and philosophical roots rather than just being an issue with social media.

---

[195] Screenshot 5 in Appendix B.
[196] See Appendix A.

The prevailing anthropology asserted by our rationalist and intellectualist model suggests that we are a sum of what we think. This would account for the greater proportion of intellectual debates surrounding social issues, a brain-storming of ideas for real-world problems or a sign-posting to content from other media platforms. However, what would it look like if our online communities "instead of starting from the assumption that human beings are thinking beings... started from the conviction that human beings are first and foremost lovers"?[197] What would it feel like to participate in a virtual community where people could bare their chests and be vulnerable with one another?

One example of posts which are personal and vulnerable are prayer requests from members. In the Catholic Geeks community, one user requested prayer for her grandmother who is going through the late stages of Alzheimer's and is experiencing heart difficulties.[198] Curiously, the OP acknow-ledges that she rarely posts in the group and usually only comments on the posts of others. This post garnered a whopping total of 120 comments which mainly promised prayers and love through the "heart" or "praying hands" emojis.[199] There were also other users who empathized with the OP and shared their struggles with ailing parents/grandparents as well, creating a sense of solidarity with the OP.

I am part of an online private group (read: not open to the general public) called Sister Lucia Dos Santo[200] that was started by my classmate at Holy Apostles College & Seminary.

---

[197] Smith, J., *You Are What You Love: The Spiritual Power of Habit.* (Grand Rapids, MI: Brazos Press, 2016), 7.

[198] Ximena Morales, 29 November 2019, 19:21.

[199] Screenshot 6 in Appendix B.

[200] See Appendix A.

Perhaps due to the small size of the community or that the members generally know one another, the posts are more personal. People share about their health and spiritual struggles and other members rally around in prayer, support and encouragement. While I have not met any of the other members of the group including my classmate, I feel close and supported by this community of individuals. Indeed, what if online communities conceived of individuals primarily as lovers and relational beings instead of rational thinkers? It promises to offer a glimpse here on earth of the heavenly communion promised to us!

Infractions and Moderation

Misbehavior in online communities will be analyzed along the lines of rule-breaking, rule-making and rule-enforcement since "misbehavior clearly involves the breaking of rules, while the regulation of misbehavior... involves the making of rules and their enforcement."[201] Further, contrary to common sense, rule-breaking tends to precede rule-making since the rules develop as a reactive way of dealing with behavior which "transgresses previously unstated, informal rules and norms."[202]

Further, given the varying needs of online communities, there would be varying degrees on which behaviors are normative and which are not. In practice, having a "rough consensus about normative behaviors can help the community achieve its mission."[203] Online Catholic communities dedicated to building communion and galvanizing zeal for evangelization must guard against the things which are

---

[201] *Misbehavior in Cyber Places*, 187.
[202] *Misbehavior in Cyber Places*, 188.
[203] *Building Successful Online Communities*, 111.

antithetical to community, i.e. behavior that reduces "human beings to units of consumption or competing interest groups or manipulate(s) viewers... as mere ciphers from whom some advantage is sought."[204]

Catholic Geeks has an extensive rules and moderation policy that is updated periodically and is bespoke for the group ("CG Rules"). The Catholic Geeks community aims to "bring Catholics together in charity and civility to share... mutual geeky interests and (to) grow in... faith."[205] The rules are provided in both a condensed and full version for the benefit of those who have shorter attention spans on a new media platform like Facebook. These are also pinned on every page of the online group so that it is clearly visible to all members.

The CG Rules clearly delineate the scope of the community's mission. One rule stipulates that theology is allowed, to an extent. Since the group celebrates geeky interests, most of the theological debates are redirected to an affiliated group called Catholic Theology Geeks. Further, as a community that celebrates geeky interests, fandom bashing is also not allowed. Thus, for example, if someone expresses that he or she is a fan of the Harry Potter series, this community will not tolerate members who slam them for being magic idolaters.

Further, the CG Rules have specific provisions regarding the issue of the Church scandal involving clerical sex abuse. While it recognizes that there is a space to discuss the Church's difficulties with the issue, the administrators have

---

[204] Pope John Paul II, "Sustained by the Spirit, Communicate Hope," Message for the 32nd World Communications Day, 24 May 1998.

[205]

https://www.facebook.com/groups/Catholicgeeks/permalink/2370340586540115/

decided to only allow prayers (without details) about the sex abuse crisis and again have redirected venting and political debates to its affiliated community, Catholic Theology Geeks. Again, this reaffirms the community's identity as being a safe family-friendly space to share geeky interests rather than to engage in political debate.

Finally, unauthorized promotional posts on the Catholic Geek community are considered spam and are deleted. The community administrators only allow for promotion allowed in the "Member Businesses" permanent document. The moderators also run a monthly special event called the "Shameless Plug" where members can advertise or promote anything that is not contrary to the rules. These includes fundraisers, petitions, YouTube channels and blogs.[206]

Given the nature of moderation, i.e., that offensive posts are removed, it is very challenging to observe whether the CG Rules are enforced. However, a survey of the posts from October to November 2019 did not show any content or interactions which flouted the CG Rules. Occasionally, a link to a YouTube video or an event invitation was spotted. While these were technically promotional material, members were sharing resources and events that were inspiring to them rather than just promoting their own ministries or services. As far as is observable from just the community discussion posts, the moderation seems to be adequately enforced.

While the TOB Institute Community and the CatholicNYC group do not have any community guidelines, Catholics in Singapore! has a short moderation guideline. The Singapore community reminds all members that the group is meant to edify its members and to help them "grow in faith and in virtue" and that all people of the receiving end of comments are children of God. Thus, comments which go against this

---

[206] Screenshot 7 in Appendix B

mission are deleted, and permanent offenders are removed from the group. The moderators also encourage a whistle-blowing policy for members to report content infringement from their fellow members.

The Singapore community also provides a simple principle in correcting another person in the online community: "If we feel like we need to correct someone, let's remember that friendship rather than rebuke is our best weapon." This parallels the Catholic practice of fraternal correction, a spiritual work of mercy which admonishes another to "protect him from sin or to induce him to give up sin."[207] Fraternal correction is to be juxtaposed with paternal correction which is administered by a superior in the capacity of father. Thus, it is an act of charity that we bear out of brotherhood with one another, i.e., that we correct in the light of our friendship with one another.

In a recent post on Catholics in Singapore!, one member lamented how she felt targeted by other members of the congregation when her baby started crying during mass.[208] While some members identified with her struggle and encouraged her to continue bringing her children to mass, there were other members who felt like she should take her crying children outside of the main worship hall.[209]

Nevertheless, despite the disagreement in perspective in what could potentially be a very emotional topic, the debates were not polarizing nor was the OP or any other users put down for their views. While polite disagreement is different from correcting others, being able to express a contrary

---

[207] Connell, FJ, "Correction, Fraternal", in *New Catholic Encyclopedia*, (2nd ed, Detroit, MI: Gale, 2003), 273

[208] Leong Yiing, 1 December 2019, 08:50

[209] Screenshot 8 in Appendix B

opinion in a civilized way provides the bedrock for being able to correct amicably.

If the moderators were interviewed or the communities were monitored in real time over a longer period, we would better understand if the rules were formed as a means of concretizing informal norms. With a limited scope, this paper is unable to comment on the retroactivity of rules.

Renewing Catholic Culture

While we can endlessly analyze the online Catholic communities across various yardsticks or identify posts which show best practices, it is perhaps more useful to ground the building of authentic online Catholics communities based on renewing Catholic culture through new evangelization.

Since social media is a "digital battleground" and simultaneously the "digital Church, forum, and classroom,"[210] we ought to exercise proper stewardship over our social network platforms so that they are consonant with the dignity of the human person and man's need to be in communion with others. Our participation and involvement with online Catholic communities become imbued with the mission of safeguarding our neighbor's dignity while fostering a place where the Gospel values take deeper root.

Further, through the Church's dialogue with culture through the process of inculturation, the person of Jesus can be incarnate in our way of life, including the online communities to which we belong. A means of renewing the culture of inauthenticity and polarizing diatribe on the internet is to introduce a Catholicity in our online communities; that we

---

[210] *Social Media Magisterium*, Loc 1624.

promote a culture "informed by revelation, robed in flesh, enriched through time."[211]

One practical way to reinvigorate our online Catholic communities and to align it with the mission of evangelization and communion of the Church is to be intentional with our engagement with social media and one another. Concretely, the "right response in each and every situation is pause,"[212] to consider what we are doing and how it affects our neighbours, Catholics and non-Catholics in the group and the wider internet population.

Conclusion

Since we find ourselves in this digital age, social media and online communities are here to stay, whether the Catholic Church likes it or not. Indeed, if the Church chooses not to engage with the digital natives, i.e., those who are fluent with social media, they may then form other communities online which may not be consonant with the Church's teachings. Thus, there is great evangelizing potential if the Church chooses to use social media as a platform for building communion.

This chapter has sought to identify best practices in online Catholic communities in relation to content, interactions and moderation for bad behavior. Generally, we find greater participation when content is member-produced rather than broadcasted by the administrators. We also observed how a tailored and strong moderation policy provides a safe environment for members to share their views online.

---

[211] Topping, R., *Rebuilding Catholic Culture: How the Catechism Can Shape Our Common Life*. (Manchester, NH: Sophia Institute Press, 2012), 227.

[212] *Social Media Magisterium*, Loc 1652.

Admittedly, it is difficult to generalize recommendations across various online Catholic communities given the varying needs. Perhaps it is more useful to recognize that social networks like Facebook are "nourished by aspirations rooted in the human heart"[213] and that we can foster a spirit of communion as we encounter the truth that we are created for love and for relationships, that as we are created in the image of God who delights in revealing Himself, we too delight in being vulnerable with one another in a safe space.

Afternote: Best Practices for a Flourishing Online Catholic Community

Pareto Principle in Content Management. Motivate most users to generate content and keep event information to roughly 20% of content.

Customized and Robust Moderation Policy. Develop and regularly maintain a moderation policy that fits the needs and sophistication of the group. Consider the mission of the group, what types of infractions are likely to occur and how best to manage them.

---

[213] Pope Benedict XVI, "Social Networks: Portals of Truth and Faith; New Spaces for Evangelization", Message for the 47th World Communications Day, 12 May 2013

Appendix A – List of online Catholic Communities

Catholic Geek: https://www.facebook.com/groups/Catholicgeeks/
CatholicMom.com:
https://www.facebook.com/CatholicMomCommunity/
CatholicNYC: https://www.facebook.com/CatholicNYC/
Catholics in Singapore! (private):
https://www.facebook.com/groups/2345489644/
Sr Lucia Dos Santos Prayer Group (private):
https://www.facebook.com/groups/343309969845865/
TOB Institute Community (private):
https://www.facebook.com/groups/tobinstitute/

Appendix B – Screenshots from Facebook communities

Screenshot 1: Event on CatholicNYC

Screenshot 2 – Meet the Staff in CatholicNYC

CatholicNYC
18 November at 10:25 · 🌐

This is "Meet the Staff" week! Each day we will highlight a different staff member of the Office of Young Adult Outreach. Meet the Staff: Colin Nykaza is the Director of the Office of Young Adult Outreach. Colin loves sports (especially rooting for the Mets!), spending time with his nieces and nephew, and spreading the message of Divine Mercy and Marian Consecration.

👍❤️ 32   1 comment 2 shares

Screenshot 3: Digital Prayer Cards on CatholicNYC

"At that time Michael, the great prince who protects your people, will arise. There will be a time of distress such as has not happened from the beginning of nations until then."
Daniel 12:1

👍❤️ 30   2 comments 4 shares

👍 Like   💬 Comment   ↪ Share   ▾

Most relevant ▾

Write a comment...   ☺ GIF 🏷️

Lorraine Freeman PRAYERS
Like · Reply · 3w

Betty Guman
✝️ ℹ️ 💜
Like · Reply · 3w

Screenshot 4 —Member reaching out for help on TOB Institute Community

Michelle Piccolo
9 November at 12:17

Question for the men: I'm giving a talk to a small group of men (guessing mostly are married and middle aged and older, but there may be some singles) to discuss "some aspect of TOB." What do you think is one of the most important parts of TOB for men to know?

👍❤️ 3   11 comments

👍 Like   💬 Comment

Hudson Byblow That chastity is a virtue for all and that virtue begets virtue and that it also begets joy- and thus increases joy in single or married life.

Like · Reply · 3w   👍❤️ 7

Toni Mihalik-Esposito I think too that we need to talk about st Joseph and how he was the protector of mary. We as women have come to fear men as a predator. But we should feel safe around them as a protector

Like · Reply · 3w   👍❤️ 2

> **Michelle Piccolo** Protector, not predator, YES!!! Oh that's a good one, thanks girl!
>
> Like · Reply · 3w   ❤️ 1

Scott Byers Ephesians 5!!!
Women's submission is not to men's domination, but to men's service to their spouse and family.
It's a mutual submission out of reverence for Christ! ... See more

Like · Reply · 3w   👍❤️ 8

> **Michelle Piccolo** YES!
>
> Like · Reply · 3w

Lauretta Robbins Sesock When my husband and I presented TOB the most lively discussion from the men always centered around contraception and the necessity of mastery of self.

Like · Reply · 3w   👍 3

Michelle Piccolo Thanks everyone for your suggestions! I recorded (audio only but I attached the slides) the presentation to the men here for those who would like to listen:
https://vimeo.com/373182870 They loved it! P.S. you do have to turn up the volume quite a bit, sorry about that.

he body, in fact, and only
the body, is capable of   𝑖
VIMEO.COM

138

Screenshot 5 – Responses to post about Vietnamese trafficking on Catholics in Singapore!

Michele Hendroff 🐿 Something similar happened in 2000 where almost 60 bodies of Chinese nationals were found in a refrigerated truck. Who doesn't want a better life? How would the victims or their families know they were dealing with traffickers? The Chuch teaches its congregation to lead a good and godly life, it can't possibly be one step ahead of every evil present.

Like · Reply · 6d

Catherine Koh The lure of money and better life...that makes even folks in developed country like Singapore to fall prey to cyber scams and losing thousand of dollars, while folks in developing countries to lose more 🙈

Like · Reply · 6d

Leong Anna It's almost impossible to tell the poor and hungry to eat healthy.

Like · Reply · 6d

👍 1

Szeache 🐿 These villagers are from the billionaire villages in Vietnam. They send their children overseas so they can repatriate their incomes home, so the families can build themselves bungalows.

Like · Reply · 6d · Edited

Martinaann Choy May they rest in peace.
May GOD give strength to the families.

Like · Reply · 6d

Screenshot 6 – Responses to prayer request on Catholic Geeks

Maria Stormont Amen
Like · Reply · 5d ♥ 1

Lyn Moody

♥ 1

Like · Reply · 5d

Lyn Moody My prayers for your Grandmother. My mother is also suffering from Alzheimer's. It has been a very difficult thing for her and us. May God bring blessings to your grandmother.

Like · Reply · 5d ♥ 1

> **Ximena Morales** Thank you so much 🖤 May God bless your mother as well and bring her health and happiness above anything 🙂
>
> Like · Reply · 4d ♥ 1

Christine Benton Marzo

Like · Reply · 5d ♥ 1

Donna Cosenza

🙏 ❤️ ♥ 1

Like · Reply · 5d

Tommie Muñoz Tran May God continue to give you strength for this difficult journey.
Praying for your grandmother Ines that she have a peaceful passing if that is God's will for her. 🙏 ❤️

Like · Reply · 5d ♥ 1

Screenshot 7: Shameless Plug on Catholic Geeks for ad-hoc promotional content

Catholic Geeks ▶ Catholic Geeks
2 July · 🌐

👍 Like Page   •••

July 2019 Shameless Plug Post -
(Here's your chance to promote stuff! 🙂)

Most know the rules, but here is a quick recap.

1. Promotional content will be limited only to comments on this thread. This includes charities, businesses, fundraisers, surveys, blogs, websites, non-CG-affiliated groups, pages, youTube channels, etc.
2. No PMing members without permission.
3. No heretical nonsense.
4. All the group rules apply here, too. You may always access the abbreviated rules through the "About" tab at the top of the group wall. In addition, a more detailed rules post is pinned to the top of the announcements section, near the top of the group wall.

Please note all GoFundMe or donation requests have not been verified by the admin team and if you give you do so at your own risk.

AND NOW A SHAMELESS PLUG:

😀👍💬 17   57 comments

👍 Like   💬 Comment

Screenshot 8: Responses to post on children crying during mass on Catholics in Singapore!

Lisa Justine When my kids were little and at mass with us, I made sure I had things to keep them occupied and took them out if they started to fuss. No one stopped me from bringing the kids to mass but I also did not let my kids' ruckus impact other people's need f...
See more

Like · Reply · 3d · Edited   11

Alicia Loo Don't be discouraged! When my kids were small, I used to sob at the pews too... but I learnt through the years, to lift it up to Jesus! When I hear the crying sounds, I would pray for baby, then mommy or daddy and then for the people around cos only J...
See more

Like · Reply · 3d · Edited   3

Shawn Xue I've developed the ability to tune most crying out. I now mostly only get distracted when the parents tell their babies to stop; in full sentences. Don't stop bringing your kids to mass. Gotta start them young.

Like · Reply · 3d   4

Elizabeth Chow Some parents bring their kids to church like they are going on a picnic. They bring their cookies, their drinks and toys to play. I feel that parents need to teach their children that they are going to church and not a picnic.
I used to bring a picture book on mass which were sold in the church's bookshop and explain the different parts of the mass to them. In those days nobody ever brought food into a church for the kids. Milk was fed to the kids outside of the church.

Like · Reply · 3d   7

Florence Heng Now i also noticed some parents preen their daughters' hair while the homily is on. OMG....

Like · Reply · 3d

Anastasia WS I'm of the camp that if my kid is crying, (s)he needs to be taken out of the church till (s)he calms down. If you can't hear the sermon or prayers over the crying, the kid's too loud.

Like · Reply · 3d   3

Annette Miranda I feel it is ok to have your children with you during mass, most priests are happy about it, although they can be a bit noisy to others. Learning the skills to manage them takes time and effort, but your patience will be so rewarding in the end when they all grow up to be God-fearing adults.

Like · Reply · 3d   2

10

Love and Responsibility: The Personalization Principle in Cyberspace[214]

Dr. Sebastian Mahfood, OP
Professor of Interdisciplinary Studies
Holy Apostles College & Seminary, Cromwell, CT

The nature of technology, according to Marshall McLuhan in *Understanding Media: The Extensions of Man*, is found in its being an extension of the human person, and the nature of cyberspace is a forum within which human persons are enabled to extend themselves to one another. It is for this reason that our use of Web sites, blogs, videos, and other interactive media we find online is a *personal encounter* with another in the sense that real persons are involved on the other end of any connection. This idea may be counter intuitive to our experience and training because we are not used to considering, say, images we find on Google's image search as extensions of the person who posted them. We tend to think of them in terms of orphaned artifacts laying around

---

[214] This chapter was first published in *Seminary Journal* (spring 2008) and is republished here with a few updates by permission of the *Journal's* General Editor.

waiting for someone to make use of them.

Based on this perception of cyberspace as a vast treasure-trove of materials ready for appropriation, we often make use of whatever we find without seeking the person who has extended him- or herself toward us. This culture of use has the potential to color all our online social relationships, not only those in which we interact with synchronously or asynchronously with other human persons but also those in which we interact only with the materials others leave behind them. We would, otherwise, grow accustomed to viewing those we meet through the communicative media of cyberspace as objects, as means to our own ends. Before we can envision others within cyberspace as authentically human, that is, as people who ought to be valued because of their inherent dignity as both individual substances of a rational nature and as beings created in the image and likeness of God, we have to develop an understanding of cyberspace as media that enables in all online encounters our relationships, however ephemeral, with other persons.

All human relationships, because of the acts that transpire within them, carry moral weight. This moral weight is, moreover, measurable in terms of goodness and evil. John Paul II writes in Veritatis Splendor that "[h]uman acts are moral acts because they express and determine the goodness or evil of the individual who performs them" (§ 71). Our engaging others in virtual relationships should, therefore, be an activity that builds upon our application of the traditional moral norms precisely because we cannot compartmentalize our morality into that which we pursue face-to-face and that which we pursue in cyberspace. Relationships among human persons are always interpersonal even if they are "site-specific." Without this understanding that our communicative partners, even in the form of our interacting with the extensions of themselves they generate in cyberspace, are persons, it is too easy for us to view those we encounter in cyberspace merely as means to our own ends without responding to them properly through love.

The culture of use that is created when we inappropriately handle the extensions of other persons is antithetical to the culture of love we seminary formators endeavor to engender within the future priesthood. That most of us have never been trained in the grammar of cyberspace sufficient for us to meaningfully evaluate its moral norms is a situation we have to correct if we are to have any credibility with our students who arrive with a great deal of technological savvy but little grasp of the role technology ought to play in service to the Church. It is for this reason that we have to meaningfully address the nature of cyberspace as communicative media within all programs of pastoral formation, a mandate given to us in 2005 by Pope John Paul II in his apostolic letter *The Rapid Development*, which requires us to prioritize making ourselves conversant with the relational nature of cyberspace so that we can pass this understanding on to our students.

The Culture of Use

In *Love and Responsibility*, published almost two decades before *Theology of the Body* began making its debut in John Paul II's Wednesday conferences, Karol Wojtyla defined the two meanings of the word "use" predicated upon the principle that the only proper way to respond to another human person is through love. In a revision of Immanuel Kant's moral imperative, Wojtyla writes that "whenever a person is the object of your activity, remember that you may not treat that person as only the means to an end, as an instrument, but must allow for the fact that he or she, too, has, or at least should have, distinct personal ends. This principle lies at the basis of all the human freedoms, properly understood, and especially freedom of conscience" (28). A human person, he emphasizes, cannot be a tool or an object acted upon entirely for another's selfish benefit, for "use" in such a way is antithetical to love.

Wojtyla's two definitions are significant because he is speaking in terms of sexual relationships. In his first definition, Wojtyla identifies "use" as the objectification of someone while in the second definition he identifies "use" as the enjoyment of someone consonant to the purpose for which the human person was created. Applying this principle to cyberspace, we might legitimately argue that the artifacts we find online are not the same as the human person who posts them, but they are extensions of the human person, representing that person in cyberspace. We might also legitimately argue that publication of any images, not just those of the body, presupposes permission for use by others, that by "virtue" of one's extending him- or herself into the public realm, he or she is transformed from a subject to an object. To argue such, though, would be to argue for the objectification of the human person through his extensions, and that is something we would do better to avoid.

To prevent our objectifying others based on our ability to appropriate their extensions to our own use, we should personalize the objects we find online; that is to say, we should become consciously aware that the efficient cause of every text, image, audio, or video is a human person. Such a personalization of that which we find in cyberspace affects our entire view of how online interactions should be conducted. Of these online interactions, two kinds exist, the asynchronous and the synchronous, respectively, those in which we encounter artifacts of others, such as text postings, images, audios, or videos (otherwise known as asynchronous interaction), and those in which we encounter the real-time presence of others interacting with us, such as textual, audio, or video chats (otherwise known as synchronous interaction). The fact that the first type is often not personalized leads ironically to the second type's depersonalization, for when we come to think of a posted video as an object without a human person behind it, the objectification of a person engaging in live video chat is not far behind.

Virtual interactions have real consequences, but it is not the consequences alone that determine the good or evil of our actions; rather, our morality as Catholics involves, of course, an intimacy in relations among our intentions, the object of our focus, and the circumstances. All three of these sources of morality must be good for the activity itself to be good, and if the good is present in only one or two of them, then the activity itself is not good. Before we even reach the consequences, then, the good inherent within any given act is already measurable by the goodness present in each of the three fonts. Our personalization of our virtual encounters, as a result, provides the key we need to discern whether what we do online is morally licit; it opens to us, therefore, the tools we need to discern whether our designs in cyberspace are to be engaged or to be avoided, for we begin to consciously apply to our relationships online the formula we use in our relationships face-to-face.

When we enter cyberspace, we usually do so with our own ends in mind, and we use the computer as a means to accomplishing those ends—either to be productive in some act or some interaction or to enrich ourselves or others in some form of study or entertainment. In the first case, productivity is never an end in and of itself but the pursuit of an end beyond itself that addresses the question of "what for?" In the second case, enrichment is considered its own end in our occupying ourselves for a time in a pleasurable activity. Wojtyla writes in Love and Responsibility, that "[m]an, precisely because he has the power to reason, can, in his actions, not only clearly distinguish pleasure from its opposite, but can also isolate it, so to speak, and treat it as a distinct aim of his activity. His actions are then shaped only with a view to the pleasure he wishes to obtain, or the pain he wishes to avoid" (33). Such pursuits in media that facilitate the pleasurable stimulation of our senses can end in the brief satisfaction of the appetite but may foster an addiction to instant gratification that is intrinsically unhealthy.

Far from being a Hobbesian or even Utilitarian view of the pleasure principle, Wojtyla's explanation that pleasure can be treated as a distinct aim of a human person establishes the necessity for a personalistic approach to online interactions. He cautions that "[i]f actions involving a person...are shaped exclusively or primarily with [another's own pleasure] in view, then that person will become only the means to an end—and 'use' in its second meaning (= enjoy) represents, as we see, a particular variant of 'use' in its first meaning" (33). (In quoting the text, I have ellipsied out an important qualify-cation "of the opposite sex" since I am applying the principle here more broadly than the sexual relationships that I will talk about in the next section.) Our accomplishing our ends via cyberspace, then, requires as much an examination of our intentions, the object of our pursuit, and the circumstances within which we pursue it as does the effort to accomplish our ends in face-to-face communities. In fact, it requires more of an examination since it does not come naturally to us to see everything in cyberspace as an extension of a particular human person and realize that real individuals with eternal destinies like ours are behind them.

The Nature of Online Pornography

The most obvious instance of the objectification of the human person is found in online pornography where persons are not valued as ends but used merely as means for the gratification of another. For this reason, the nature of online pornography is often considered to rest in its ubiquity, a fact that enables the anonymity of the end-user, thus fostering hedonism and even solipsism. While this is true for those who participate in online pornography, the activity of anonymous engagement does not define the nature of online pornography, which is itself derived from the nature of cyberspace as a forum in which human persons are enabled to extend themselves to one another. The nature of online pornography, rather, lies in its interactivity, or the ability of users to become the pornography they want to see in the world. This is accomplished either vicariously through offscreen participation in one's own genital manipulation (and, almost as bad, curiosity that leads one into unchaste thoughts or activities) or publicly through onscreen participation in the genital manipulation of one another. In such activity, it is not the human person who is desired but the various fetishes and body parts by which voyeurs are attracted.

Participation is key because every encounter in cyberspace under the personalization principle is an encounter with another person even in cases where that other person's knowledge or explicit consent is lacking. In the first case, the participant relies upon the asynchronicity of cyberspace—on posted texts, images, audios, or videos that represent past activity that is not happening live. Devices exist that can be attached to the body and connected through the USB port to external stimulators of various sorts controlled by the voyeur's mouse. In the second case, the participant relies upon the synchronicity of cyberspace—on texts, images, audios, or videos that represent present activity that is happening live. The participant can engage other persons in the live act through the use of an instant chat forum, an audio hookup, or a Web-camera video feed. Sufficient anonymity is found not in the hiding of one's face, which some adult chat rooms require to be shown to guard against participation by minors, but in the masking of one's real name. Video feeds can, nonetheless, be archived by anyone with a screen recorder for any given use beyond the immediate context of situation in which the film was captured. The personalization principle quickly loses its application in a context where everyone involved in the attempt to form a community of voyeurs is ultimately exposed as seeking only selfish ends.

That the end of online pornography, also known as cybersex when it is used for genital manipulation intended to lead to orgasm, is not morally licit lies in the fact that it does not fulfill the criteria of the good as expressed in any of the three fonts. In the case of the intention, the participant engages in a form of solipsism that has as its ultimate end self-gratification at the expense of another, the expense being the use of another as merely a means. In the case of the object, the persons involved on either end of the communicative channel are not interested in the personhood of the other but in the other's manipulation of his or her own parts, an act that is designed to lead to unchaste thoughts or actions on the part of the viewer. Wojtyla is instructive as he writes,

> A person of the opposite sex [or of the same sex, for that matter] cannot be for another person only the means to an end—in this case sexual pleasure or delight. The belief that a human being is a person leads to the acceptance of the postulate that enjoyment must be subordinated to love. "Use," not only in the first, broader and more objective, meaning, but also in its second, narrower, more subjective meaning (for the experience of pleasure is by its nature subjective) can be raised to the level appropriate to an interpersonal

relationship only by love. (34, brackets mine)

Cybersex, of course, has nothing to do with love and everything to do with lust precisely because of its object, which is either a desire for control of someone else on the part of the exhibitionist or a desire for gratification of the self on the part of the voyeur. No middle ground exists.

The claim is often made by participants that through cybersex they can express themselves more meaningfully to others than they would otherwise because of geographical distance or temporal unavailability. This idea of a meaningful presence, though, cannot be an expression of love unless the Holy Spirit is enabled to participate fully in the act through covenantal, embodied sex that does not seek to impede conception. "[T]he science of ethics…whenever questions of sexual morality arise," Wojtyla writes, consequently must "distinguish very carefully between whatever shows 'loving kindness', and whatever shows not that but the intention to 'use' a person even when it disguises itself as love, and seeks to legitimate itself under that name" (34). It is in this regard that the third font can be addressed, for the circumstances of cybersex are never conducive to the pursuit of the good for the very reason that the Holy Spirit never receives an invitation to participate. Taught in this light, the argument that lust is legitimate, especially when it affects no one but the voyeur, quickly loses the glamour of its appeal.

The Seminary as Guardian of Virtue

That which is viewed or done in cyberspace is wished for by the viewer, by the acting person, who confirms the existence of the kind of real-space world he or she would like to see come into being. Such a world, though, is not one that is conducive to authentic human relationships even in cyber-

space, a venue in which all activities are perceived as inno-
cuous by those who have no real understanding of the nature
of communicative media. If a person does not come to realize
the value of a human being in cyberspace, moreover, then it is
a sign that the person does not really grasp the value of the
human person in real space. Seminary formators who rightly
consider the pursuit of online pornography as not only
detrimental to the pursuit of chastity, but also a sign of
spiritual and social corruption that threatens the foundations
of charity upon which all things rest should, consequently,
apply this understanding of the personalization principle to
other forms of relational activity where the morality of the act
is less certain because of the nature of the media involved.
Because we will not always be immediately present in the lives
of our seminarians, it is more important for us to teach
morality than it is for us to enforce obedience, though the
latter often supports the former, and this teaching has to
stretch beyond the sexual in terms of human relationships to
reach even the mundane matters of, for instance, copyright
awareness in the creation of PowerPoint slideshows for use in
class presentations and, say, eventual posting online.

To address human formation concerns and demonstrate
the accountability of our administrators, many seminaries
already have in place on their networks some kind of
monitoring software that blocks access to rated sites. Most
software, however, even if it does discriminate in blocking
rated Web sites does not discriminate in blocking rated
images or videos within a Google search of sexual terms, nor
does it record the search terms within the monitoring
reports.[215] The stress on morality becomes key in circum-

---

[215] A phenomenon that has arisen since the original publication
of this article is the rise in cell phone ownership among
seminarians, which enables them to connect to the internet through

stances like these, and it is simplified through the advancement of the personalization principle. As the Pontifical Council for Social Communications writes in its 2000 publication titled "Ethics in Communications," "the contents of the countless choices made by all these people concerning the media are different from group to group and individual to individual, but the choices all have ethical weight and are subject to ethical evaluation. To choose rightly, those choosing need to 'know the principles of the moral order and apply them faithfully' (*Inter Mirifica*, 4)" (§ 4).

The principles of the moral order rest in personalization, so it makes sense that we would apply such a concept to the extensions of the human person as a way of demonstrating that such extensions always point back to the person from whom they came in a way analogous to our knowing the person of God through his effects. Because the personalization principle pervades every encounter with others in cyberspace through its dealings with all media as extensions of human persons, it is incumbent on seminary formators to develop their own understanding of the nature of cyberspace before they endeavor to teach it to others. John Paul II summoned the whole Catholic Church in "Internet: A New Forum for Proclaiming the Gospel" (2002) "to put out into the deep of the Net, so that now as in the past the great engagement of the Gospel and culture may show to the world 'the glory of God on the face of Christ' (2 Cor 4:6)" because "if there is no room for Christ, there is no room for man" (§ 9).

We seminary formators cannot do this, however, if we are ourselves unaware of what it means for us to do so. Some instruction is provided in two documents also published in 2002 by the Pontifical Council for Social Communications at

---

their phone plans and bypass entirely the school's monitoring system.

John Paul II's request. These are "Church and Internet" and "Ethics in Internet". The former speaks to the role of the Internet in the life of the Church, stating meaningfully that "[h]anging back timidly from fear of technology or for some other reason is not acceptable" ("Church and Internet," § 10), and the latter "emphasize[s] that the Catholic Church, along with other religious bodies, should have a visible, active presence on the Internet and be a partner in the public dialogue about its development" ("Ethics in Internet," § 18). If formators read only these two documents, they will find themselves pointed in the direction of other magisterial teachings on the nature and use of communicative media—namely, *Miranda Prorsus* (1957), *Inter Mirifica* (1963), *Redemptoris Missio* (1990), *Aetatis Novae* (1992), "Ethics in Communications" (2000), and the various messages of Popes Paul VI, John Paul II, Benedict XVI, and Francis on the World Communications Days. In developing some methods by which to address these mandates, we not only do ourselves a service, but we also provide a structure within which we may train future priests who may currently have technological habits at variance with the moral teachings of the Church due to their simply being unaware of the applications of the personalization principle within cyberspace.

Sebastian Mahfood, Ph.D., is Vice President and Professor of Interdisciplinary Studies at Holy Apostles College & Seminary in Cromwell, CT. He has been actively involved since 2000 in the transformation of theological studies programs through the integration of appropriate technologies in the teaching and learning environment. He holds a Ph.D. in postcolonial literature from Saint Louis University.

MAGISTERIAL RESOURCES

**World Communications Day Messages
Popes Paul VI, John Paul II, Benedict XVI, and
Francis (1967-2020)**

1. Pope Paul VI, "Church and Social Communication: First World Communication Day," 7 May 1967. Available online at http://w2.vatican.va/content/paul-vi/en/messages/communications/documents/hf_p-vi_mes_19670507_i-com-day.html
2. Pope Paul VI, "Social Communications and the Development of Nations," 26 March 1968. Available online at http://w2.vatican.va/content/paul-vi/en/messages/communications/documents/hf_p-vi_mes_19680326_ii-com-day.html
3. Pope Paul VI, "Social Communications and the Family," 7 April 1969. Available online at http://w2.vatican.va/content/paul-vi/en/messages/communications/documents/hf_p-vi_mes_19690407_iii-com-day.html
4. Pope Paul VI, "Social Communications and Youth," 6 April 1970. Available online at http://w2.vatican.va/content/paul-

vi/en/messages/communications/documents/hf_p-vi_mes_19700406_iv-com-day.html

5. Pope Paul VI, "The role of Communications Media in promoting unity among men," 25 March 1971. Available online at http://w2.vatican.va/content/paul-vi/en/messages/communications/documents/hf_p-vi_mes_19710325_v-com-day.html

6. Pope Paul VI, "The Media of Social Communications at the Service of Truth," 21 April 1972. Available online at http://w2.vatican.va/content/paul-vi/en/messages/communications/documents/hf_p-vi_mes_19720421_vi-com-day.html

7. Pope Paul VI, "The Mass Media and the Affirmation and Promotion of Spiritual Values," 1 May 1973. Available online at http://w2.vatican.va/content/paul-vi/en/messages/communications/documents/hf_p-vi_mes_19730501_vii-com-day.html

8. Pope Paul VI, "Social Communications and Evangelization in Today's World," 16 May 1974. Available online at http://w2.vatican.va/content/paul-vi/en/messages/communications/documents/hf_p-vi_mes_19740516_viii-com-day.html

9. Pope Paul VI, "The Mass Media and Reconciliation," 19 April 1975. Available online at http://w2.vatican.va/content/paul-vi/en/messages/communications/documents/hf_p-vi_mes_19750419_ix-com-day.html

10. Pope Paul VI, "Social Communications and the Fundamental Rights and Duties of Man," 11 April 1976. Available online at http://w2.vatican.va/content/paul-vi/en/messages/communications/documents/hf_p-vi_mes_19760411_x-com-day.html

11. Pope Paul VI, "Advertising in the Mass Media: Benefits, Dangers, Responsibilities," 12 May 1977. Available online

at http://w2.vatican.va/content/paul-vi/en/messages/communications/documents/hf_p-vi_mes_19770512_xi-com-day.html

12. Pope Paul VI, "The receiver in Social Communications; his expectations, his rights, his duties," 28 April 1978. Available online at http://w2.vatican.va/content/paul-vi/en/messages/communications/documents/hf_p-vi_mes_19780423_xii-com-day.html

13. Pope John Paul II, "Social Communications for the Development of the Child," 27 May 1979. Available online at http://w2.vatican.va/content/john-paul-ii/en/messages/communications/documents/hf_jp-ii_mes_23051979_world-communications-day.html

14. Pope John Paul II, "Social Communications and Family," 18 May 1980. Available online at http://w2.vatican.va/content/john-paul-ii/en/messages/communications/documents/hf_jp-ii_mes_01051980_world-communications-day.html

15. Pope John Paul II, "Social Communications in the Service of Responsible Human Freedom," 31 May 1981. Available online at http://w2.vatican.va/content/john-paul-ii/en/messages/communications/documents/hf_jp-ii_mes_10051981_world-communications-day.html

16. Pope John Paul II, "Social Communications and the Problems of the Elderly," 10 May 1982. Available online at http://w2.vatican.va/content/john-paul-ii/en/messages/communications/documents/hf_jp-ii_mes_10051982_world-communications-day.html

17. Pope John Paul II, "Social Communications and the Promotion of Peace," 15 May 1983. Available online at http://w2.vatican.va/content/john-paul-ii/en/messages/communications/documents/hf_jp-ii_mes_25031983_world-communications-day.html

18. Pope John Paul II, "Social Communication: Instruments of Encounter Between Faith and Culture," 3 June 1984. Available online at http://w2.vatican.va/content/john-paul-ii/en/messages/communications/documents/hf_jp-ii_mes_24051984_world-communications-day.html

19. Pope John Paul II, "Social Communications for a Christian Promotion of Youth," 19 May 1985. Available online at http://w2.vatican.va/content/john-paul-ii/en/messages/communications/documents/hf_jp-ii_mes_15041985_world-communications-day.html

20. Pope John Paul II, "Social Communications and the Christian Formation of Public Opinion," 11 May 1986. Available online at http://w2.vatican.va/content/john-paul-ii/en/messages/communications/documents/hf_jp-ii_mes_24011986_world-communications-day.html

21. Pope John Paul II, "Social Communications at the Service of Justice and Peace," 31 May 1987. Available online at http://w2.vatican.va/content/john-paul-ii/en/messages/communications/documents/hf_jp-ii_mes_24011987_world-communications-day.html

22. Pope John Paul II, "Social Communications and the Promotion of Solidarity and Fraternity Between Peoples and Nations," 15 May 1988. Available online at http://w2.vatican.va/content/john-paul-ii/en/messages/communications/documents/hf_jp-ii_mes_24011988_world-communications-day.html

23. Pope John Paul II, "Religion in the Mass Media," 7 May 1989. Available online at http://w2.vatican.va/content/john-paul-ii/en/messages/communications/documents/hf_jp-ii_mes_24011989_world-communications-day.html

24. Pope John Paul II, "The Christian Message in a Computer Culture," 27 May 1990. Available online at http://w2.vatican.va/content/john-paul-ii/en/messages/communications/documents/hf_jp-ii_mes_24011990_world-communications-day.html

25. Pope John Paul II, "The Communications Media and the Unity and Progress of the Human Family," 12 May 1991. Available online at http://w2.vatican.va/content/john-paul-ii/en/messages/communications/documents/hf_jp-ii_mes_24011991_world-communications-day.html

26. Pope John Paul II, "The Proclamation of Christ's Message in the Communications Media," 31 May 1992. Available online at http://w2.vatican.va/content/john-paul-ii/en/messages/communications/documents/hf_jp-ii_mes_24011992_world-communications-day.html

27. Pope John Paul II, "Videocassettes and audiocassettes in the formation of culture and of conscience," 23 May 1993. Available online at http://w2.vatican.va/content/john-paul-ii/en/messages/communications/documents/hf_jp-ii_mes_24011993_world-communications-day.html

28. Pope John Paul II, "Television and family: guidelines for good viewing," 24 January 1994. Available online at http://w2.vatican.va/content/john-paul-ii/en/messages/communications/documents/hf_jp-ii_mes_24011994_world-communications-day.html

29. Pope John Paul II, "Cinema: Communicator of Culture and of Values," 28 May 1995. Available online at http://w2.vatican.va/content/john-paul-ii/en/messages/communications/documents/hf_jp-ii_mes_06011995_world-communications-day.html

30. Pope John Paul II, "The Media: Modern Forum for Promoting the Role of Women in Society," 19 May 1996.

Available online at http://w2.vatican.va/content/john-paul-ii/en/messages/communications/documents/hf_jp-ii_mes_24011996_world-communications-day.html

31. Pope John Paul II, "Communicating Jesus: The Way, the Truth and the Life," 11 May 1997. Available online at http://w2.vatican.va/content/john-paul-ii/en/messages/communications/documents/hf_jp-ii_mes_24011997_world-communications-day.html

32. Pope John Paul II, "Sustained by the Spirit, communicate hope," 24 May 1998. Available online at http://w2.vatican.va/content/john-paul-ii/en/messages/communications/documents/hf_jp-ii_mes_26011998_world-communications-day.html

33. Pope John Paul II, "Mass media: a friendly companion for those in search of the Father," 16 May 1999. Available online at http://w2.vatican.va/content/john-paul-ii/en/messages/communications/documents/hf_jp-ii_mes_24011999_world-communications-day.html

34. Pope John Paul II, "Proclaiming Christ in the Media at the Dawn of the New Millennium," 4 June 2000. Available online at http://w2.vatican.va/content/john-paul-ii/en/messages/communications/documents/hf_jp-ii_mes_20000124_world-communications-day.html

35. Pope John Paul II, "'Preach from the housetops'": The Gospel in the Age of Global Communication," 27 May 2001. Available online at http://w2.vatican.va/content/john-paul-ii/en/messages/communications/documents/hf_jp-ii_mes_20010124_world-communications-day.html

36. Pope John Paul II, "Internet: A New Forum for Proclaiming the Gospel," 12 May 2002. Available online at http://w2.vatican.va/content/john-paul-

ii/en/messages/communications/documents/hf_jp-
ii_mes_20020122_world-communications-day.html

37. Pope John Paul II, "The Communications Media at the Service of Authentic Peace in the Light of 'Pacem in Terris,'" 1 June 2003. Available online at http://w2.vatican.va/content/john-paul-ii/en/messages/communications/documents/hf_jp-ii_mes_20030124_world-communications-day.html

38. Pope John Paul II, "The Media and the Family: A Risk and a Richness," 23 May 2004. Available online at http://w2.vatican.va/content/john-paul-ii/en/messages/communications/documents/hf_jp-ii_mes_20040124_world-communications-day.html

39. Pope John Paul II, "The Communications Media: At the Service of Understanding Among Peoples," 8 May 2005. Available online at http://w2.vatican.va/content/john-paul-ii/en/messages/communications/documents/hf_jp-ii_mes_20050124_world-communications-day.html

40. Pope Benedict XVI, "The Media: A Network for Communication, Communion and Cooperation," 28 May 2006. Available online at http://w2.vatican.va/content/benedict-xvi/en/messages/communications/documents/hf_ben-xvi_mes_20060124_40th-world-communications-day.html

41. Pope Benedict XVI, "Children and the Media: A Challenge for Education," 20 May 2007. Available online at http://w2.vatican.va/content/benedict-xvi/en/messages/communications/documents/hf_ben-xvi_mes_20070124_41st-world-communications-day.html

42. Pope Benedict XVI, "The Media: At the Crossroads between Self-Promotion and Service. Searching for the

Truth in order to Share it with Others," 4 May 2008. Available online at http://w2.vatican.va/content/benedict-xvi/en/messages/communications/documents/hf_ben-xvi_mes_20080124_42nd-world-communications-day.html

43. Pope Benedict XVI, "New Technologies, New Relationships. Promoting a Culture of Respect, Dialogue and Friendship," 24 May 2009. Available online at http://w2.vatican.va/content/benedict-xvi/en/messages/communications/documents/hf_ben-xvi_mes_20090124_43rd-world-communications-day.html

44. Pope Benedict XVI, "The Priest and Pastoral Ministry in a Digital World: New Media at the Service of the Word," 16 May 2010. Available online at http://w2.vatican.va/content/benedict-xvi/en/messages/communications/documents/hf_ben-xvi_mes_20100124_44th-world-communications-day.html

45. Pope Benedict XVI, "Truth, Proclamation and Authenticity of Life in the Digital Age," June 5, 2011. Available online at http://w2.vatican.va/content/benedict-xvi/en/messages/communications/documents/hf_ben-xvi_mes_20110124_45th-world-communications-day.html

46. Pope Benedict XVI, "Silence and Word: Path of Evangelization," 20 May 2012. Available online at http://w2.vatican.va/content/benedict-xvi/en/messages/communications/documents/hf_ben-xvi_mes_20120124_46th-world-communications-day.html

47. Pope Benedict XVI, "Social Networks: portals of truth and faith; new spaces for evangelization," 12 May 2013. Available online at http://w2.vatican.va/content/benedict-xvi/en/messages/communications/documents/hf_ben-xvi_mes_20130124_47th-world-communications-day.html

48. Pope Francis, "Communication at the Service of an Authentic Culture of Encounter," 24 January 2014. Available online at http://w2.vatican.va/content/francesco/en/messages/communications/documents/papa-francesco_20140124_messaggio-comunicazioni-sociali.html

49. Pope Francis, "Communicating the Family: A Privileged Place of Encounter with the Gift of Love," 23 January 2015. Available online at http://w2.vatican.va/content/francesco/en/messages/communications/documents/papa-francesco_20150123_messaggio-comunicazioni-sociali.html

50. Pope Francis, "Communication and Mercy: A Fruitful Encounter," 24 January 2016. Available online at http://w2.vatican.va/content/francesco/en/messages/communications/documents/papa-francesco_20160124_messaggio-comunicazioni-sociali.html

51. Pope Francis, "'Fear not, for I am with you' (Is 43:5): Communicating Hope and Trust in our Time," 24 January 2017. Available online at http://w2.vatican.va/content/francesco/en/messages/communications/documents/papa-francesco_20170124_messaggio-comunicazioni-sociali.html

52. Pope Francis, "'The truth will set you free' (Jn 8:32). Fake news and journalism for peace," 24 January 2018. Available online at http://w2.vatican.va/content/francesco/en/messages/communications/documents/papa-francesco_20180124_messaggio-comunicazioni-sociali.html

53. Pope Francis, "'We are members one of another' (Eph 4,25). From social network communities to the human community," 24 January 2019. Available online at http://w2.vatican.va/content/francesco/en/messages/communications/documents/papa-francesco_20190124_messaggio-comunicazioni-sociali.html

54. Pope Francis, "'That you may tell your children and grandchildren' (Ex 10:2) Life becomes history," 24 January 2020. Available online at http://www.vatican.va/content/francesco/en/messages/communications/documents/papa-francesco_20200124_messaggio-comunicazioni-sociali.html

United States Conference of Catholic Bishops Department of Communications

1. Protocol for Catholic Media Programming and Media Outlets. August 11, 2000. http://www.usccb.org/about/communications/protocol-for-catholic-media-programming-and-media-outlets.cfm

2. Gray, Mark M., and Mary L. Gauthier. "Catholic New Media Use in the United States, 2012." Center for Applied Research in the Apostolate. Available online at http://www.usccb.org/about/communications/upload/Catholic_New_Media_Use_in_United_States_2012.pdf

3. Social Media Guidelines, June 2014. Available online at http://www.usccb.org/about/communications/social-media-guidelines.cfm

Vatican Documents

1. Pope Pius XII, *Vigilanti Cura*, June 29, 1936. Available online at http://www.vatican.va/holy_father/pius_xi/encyclicals/documents/hf_p-xi_enc_29061936_vigilanti-cura_en.html
2. Pope Pius XII, *The Ideal Film, Exhortations of His Holiness Pius XII to Representatives of the World of the Cinema*, June 21, 1955 – October 25, 1955. Available online at http://www.vatican.va/holy_father/pius_xii/apost_exhortations/documents/hf_p-xii_exh_25101955_ideal-film_en.html
3. Pope Pius XII, *Miranda Prorsus, On Cinemas, Radios and Televisions*, September 8, 1957. Available online at http://www.vatican.va/holy_father/pius_xii/encyclicals/documents/hf_p-xii_enc_08091957_miranda-prorsus_en.html
4. Pope John XXIII, *Boni Pastoris, Good Shepherd, Establishing the Pontifical Commission for Motion Pictures, Radio and Television*, February 22, 1959. Available online at http://www.vatican.va/holy_father/john_xxiii/motu_proprio/documents/hf_j-xxiii_motu-proprio_22021959_boni-pastoris_en.html
5. Pontifical Commission for the Cinema, Radio and Television, "Statute of the Vatican Film-Library," November 16, 1959. Available online at http://www.vatican.va/roman_curia/pontifical_councils

/pccs/documents/rc_pc_pccs_doc_16111959_statute-film-library_en.html

6. Vatican II, *Inter Mirifica, Decree on the Media of Social Communications*, December 4, 1963. Available online at http://www.vatican.va/roman_curia/pontifical_councils/pccs/documents/rc_pc_pccs_doc_04121963_inter-mirifica_en.html

7. Pope Paul VI, *In Fructibus Multis, Establishing the Pontifical Commission for Social Communications*, April 2, 1964. Available online at http://w2.vatican.va/content/paul-vi/it/motu_proprio/documents/hf_p-vi_motu-proprio_19640402_in-fructibus-multis.html

8. Pontifical Commission for Social Communications, "Communio et Progressio, On the Means of Social Communication written by Order of the Second Vatican Council, 23 May, 1971. Available online at http://www.vatican.va/roman_curia/pontifical_councils/pccs/documents/rc_pc_pccs_doc_23051971_communio_en.html

9. Pontifical Commission for Social Communications, "An Appeal to all Contemplative Religious," March 6, 1973. Available online at http://www.vatican.va/roman_curia/pontifical_councils/pccs/documents/rc_pc_pccs_doc_03061973_contemplative-religious_en.html

10. Pontifical Commission for Social Communications, "Guide to the Training of Future Priests concerning the Instruments of Social Communications," March 19, 1986. Available online at http://www.vatican.va/roman_curia/pontifical_councils/pccs/documents/rc_pc_pccs_doc_19031986_guide-for-future-priests_en.html

11. Pontifical Council for Social Communications, "Pornography and Violence in the Communications Media: A Pastoral Response," May 7, 1989. Available online at http://www.vatican.va/roman_curia/pontifical_councils/pccs/documents/rc_pc_pccs_doc_07051989_pornography_en.html

12. Pontifical Council for Social Communications, "Criteria for Ecumenical and Inter-Religious Cooperation in Communications," October 4, 1989. Available online at http://www.vatican.va/roman_curia/pontifical_councils/pccs/documents/rc_pc_pccs_doc_04101989_criteria_en.html

13. John Paul II, *Redemptoris Missio*, On the permanent validity of the Church's missionary mandate, December 7, 1990. Available online at http://w2.vatican.va/content/john-paul-ii/en/encyclicals/documents/hf_jp-ii_enc_07121990_redemptoris-missio.html

14. Pontifical Council for Social Communications, *Aetatis Novae, on Social Communications on the Twentieth Anniversary of Communio et Progressio*, February 22, 1992. Available online at http://www.vatican.va/roman_curia/pontifical_councils/pccs/documents/rc_pc_pccs_doc_22021992_aetatis_en.html

15. Pontifical Council for Social Communications, "100 Years of Cinema," January 1, 1996. Available online at http://www.vatican.va/roman_curia/pontifical_councils/pccs/documents/rc_pc_pccs_doc_19960101_100-cinema_en.html

16. Pontifical Council for Social Communications, "Ethics in Advertising," February 22, 1997. Available online at http://www.vatican.va/roman_curia/pontifical_councils

/pccs/documents/rc_pc_pccs_doc_22021997_ethics-in-ad_en.html

17. Pontifical Council for Social Communications, "Ethics in Communications," Vatican City, June 4, 2000. Available online at http://www.vatican.va/roman_curia/pontifical_councils/pccs/documents/rc_pc_pccs_doc_20000530_ethics-communications_en.html

18. Pontifical Council for Social Communications, "The Church and Internet," February 22, 2002. Available online at http://www.vatican.va/roman_curia/pontifical_councils/pccs/documents/rc_pc_pccs_doc_20020228_Church-internet_en.html

19. Pontifical Council for Social Communications, "Ethics in Internet," February 22, 2002. Available online at http://www.vatican.va/roman_curia/pontifical_councils/pccs/documents/rc_pc_pccs_doc_20020228_ethics-internet_en.html

20. Pope John Paul II, *The Rapid Development*, 24 January 2005. Available online at http://www.vatican.va/holy_father/john_paul_ii/apost_letters/documents/hf_jp-ii_apl_20050124_il-rapido-sviluppo_en.html

ABOUT THE EDITOR

Dr. Sebastian Mahfood, OP, is a Lay Dominican of the Chapter of the Holy Rosary in the Province of St. Albert the Great who began teaching the use of appropriate technologies in theological teaching and learning environments in 2000 at Kenrick-Glennon Seminary in St. Louis.

In 2006, Dr. Mahfood formed with a generous grant from the Wabash Center an organization called the Catholic Distance Learning Network (CDLN), which operated through the Seminary Department of the National Catholic Educational Association. He established a training program that provided certification in online teaching and learning to over a hundred seminary and theological institute faculty over the course of the next half decade and formed partnerships among two dozen of the schools to offer online courses to one another's students on the basis of reciprocity.

In 2008, Dr. Mahfood formed under the guidance of Dr. Charles Willard the Technology in Theological Education Group of the Association of Theological Schools (ATS) and provided leadership over the next four and a half years in the redevelopment of the technology standards in the accrediting body's general institutional and degree program standards, changes that resulted in the ability of theological schools to offer 100% online degree programs.

In 2011, Dr. Mahfood was named by then-President-Rector, the Very Rev. Douglas L. Mosey, CSB, the interim Director of Distance Learning at Holy Apostles College & Seminary and left Kenrick-Glennon the following year to engage the work full-time, hiring to develop and teach a new curriculum of courses many of the faculty the CDLN had earlier trained. He was also in 2011 appointed by then-President Richard Peddicord, OP, of the Aquinas Institute of Theology as a member of the school's board of trustees

Dr. Mahfood was named in 2012 by now Bishop Richard Henning the provost of the Sacred Heart Institute in Huntington, NY, and worked with the faculty of Holy Apostles to design and implement five dozen short online courses as continuing education opportunities for the priests and deacons of New York. He has served as an adjunct faculty member for both Aquinas Institute and St. Joseph's Seminary to teach the theology of the media. Over the course of the decade, he has served as a consultant to many of the US Catholic seminaries and theological schools and as a technology and online learning expert on over a dozen accreditation visits on behalf of ATS.

In 2014, Dr. Mahfood founded En Route Books and Media, LLC, available online at https://www.enroutebooks.com, as a means by which to assist the faculty of Holy Apostles College & Seminary to secure a publishing outlet for their books. Two years later, he founded WCAT Radio, available online at https://www.wcatradio.com as a means by which to assist in the marketing and promotions of the publishing house. Over a five-year period, En Route has produced 120 books in the areas of academics, popular spirituality, children's literature, and fiction. Over a three-year period, WCAT Radio has produced 68 radio programs covering the same genres, posted as a result of them over 5,000 podcasts with a listening audience of over 100,000 people in the year 2019.

Dr. Mahfood holds master's degrees in comparative literature from the University of Texas at Arlington, in philosophy and theology from Holy Apostles College & Seminary, and in educational technology from Webster University. His doctorate is in postcolonial literature and theory from Saint Louis University.

Among his publications include his book *Radical Eschatologies: Embracing the Eschaton in the Works of Ngugi wa Thiong'o, Nuruddin Farah, and Ayi Kwei Armah* (2009), a book co-authored with Dr. Ronda Chervin entitled *Catholic Realism: a Framework for the Refutation of Atheism and the Evangelization of Atheists* (2015), a book co-authored with Bishop Richard Henning of the Diocese of Rockville Centre entitled *Missionary Priests in the Homeland: Our Call to Receive* (2020) and *The Narrative Spirituality of Dante's Divine Comedy: A Hundred-Day Guided Journal* (2020). Dr. Mahfood lives in St. Louis with his wife, Dr. Stephanie Mahfood, and children, Alexander and Eva Ruth.

www.ingramcontent.com/pod-product-compliance
Lightning Source LLC
Chambersburg PA
CBHW031841090426
42741CB00005B/315